# The India Story

# The India Story

AN EPIC JOURNEY OF
DEMOCRACY AND DEVELOPMENT

## BIMAL JALAN

RUPA

First published by
Rupa Publications India Pvt. Ltd 2021
7/16, Ansari Road, Daryaganj
New Delhi 110002

*Sales Centres:*

Allahabad Bengaluru Chennai
Hyderabad Jaipur Kathmandu
Kolkata Mumbai

ISBN: 978-93-90918-81-2

First impression 2021

10 9 8 7 6 5 4 3 2 1

The moral right of the author has been asserted.

Printed by Parksons Graphics Pvt. Ltd, Mumbai.

# CONTENTS

# INTRODUCTION

Not too long ago, it had seemed that coalition governments were going to become a permanent feature of our country. Such governments, because of their very nature, pose a challenge to the reform process because consensus building between coalition partners can be a tiresome affair. So, when it seemed that coalitions were here to stay, one assumed that reforms and good governance may have to take a back seat. After all, the constant pushes and pulls of coalition partners often mean that even those behind the wheel of that coalition cannot exert their full power and potential.

Since 2014, however, the political profile of the Indian government has changed dramatically. Even while it has collaborated with other parties, the Bharatiya Janata Party has formed the government at the Centre with a full majority of its own, and that too for a second consecutive term (2019–). This puts the party in a position to launch political and economic reforms to promote India's overall national interests.

It is against this background that I present this volume, which recounts India's experiences over the years, for our learnings from the past can contribute immensely to the contemporary debate about our country's approach to its economic and political reforms in the twenty-first century.

Divided into two sections—'Learning from our Economic Past' and 'Beyond the Metrics of Economy', the essays in this volume look at the past with a view to making more informed decisions for India's future. With essays on topics ranging from economic and political reforms to good governance, it is my belief that this book will be of interest not only to policymakers and professional experts but also to an informed general reader.

◆

When we look back at events after the passage of time, we do so with the benefit of distance and wisdom. In the first section, we look at India's economic past and the learnings we can draw from it to make more informed decisions in future. It may be mentioned that some of the points given below have been highlighted in my previously published books.

While the balance-of-payments (BOP) crisis of 1990 is referred to often, India's trajectory leading to that crisis had a brief period of recovery. This was between 1976 and 1979, when India witnessed substantial improvement in its BOP and foreign exchange reserve position.[1] This brief respite came after a rather long period of almost two decades—from 1956 to 1975, which was marked by persistent BOP problems and it saw the beginnings of a slow process of industrial and import liberalization. What was disheartening, however, was that we couldn't sustain the brief period of recovery. The sharp increase in the cost of India's oil imports in the 1980s resulted in the deterioration in terms of trade and hence, the BOP problems re-emerged, which eventually culminated in the BOP crisis of 1990.

Understanding why we failed then can offer us great lessons to strengthen the future viability of BOP. And this is exactly what the chapter on 'Towards a Balance of Payments Viability' aims to do. Highlighting certain issues in India's trade policy, in this chapter, I talk about three important policy priorities: (i) the need

---

[1]'India's BOP Performance: Balance of Payment versus Balance of Trade, Current Account versus Capital Account,' Civilsdaily, 22 September 2017, https://www.civilsdaily.com/indias-bop-performance-balance-of-payment-versus-balance-of-trade-current-account-versus-capital-account/, accessed 16 March 2021.

to reduce the government's fiscal deficit; (ii) the need for certain modifications in the import-export policy; and (iii) the need of policy for commercial borrowings.

The BOP crisis of 1990 indeed triggered the need for India to look at things differently. In my next essay, titled 'After the Crisis: The Need for a New Strategy', I highlight India's prospects and the need for a new strategy to overcome the problems faced by the country in accelerating the GDP growth rate and alleviating poverty. This essay emphasizes the need for restoration of macroeconomic stability, which is crucial for investment and growth. Important areas where there is immense potential for further investment include energy efficiency and a well-functioning financial system, with further rationalization of interest rate, special deduction in corporate tax rates for investments in new plants, liberalization of foreign investments, and improvement in the operational efficiency of infrastructure and the public sector. The key to higher economic growth is an efficient investment policy, and this is where India should concentrate in the future.

In the wake of the acute economic crisis of 1990–91, the new government of the time had also launched a series of economic reforms in 1991. In the chapter 'Trade, Investments and Capital Flows', we look at India's position with regard to these three sectors and the action taken by the government to liberalize import-export trade, domestic investment policies and capital flows from abroad. The measures introduced by the government in these sectors were wide-ranging and led to increased competition in areas earlier dominated by monolithic public-sector enterprises. Our foreign exchange position was also stabilized, which further increased the growth potential of the economy and reduced poverty levels. Against this backdrop, this essay deals with certain critical policy initiatives that may be taken by the government to achieve political consensus across different political parties and regions. In the light of experience of other developing countries in respect of their successes and failures in achieving higher growth rates, it was considered desirable, in future, to learn from international experience, and work towards a more prosperous and poverty-free India in the twenty-first century. Since income disparities are still deeply entrenched in our society, India needs to make conscious and consistent efforts to stay on track in its efforts towards this important goal.

In a globalized world, where countries and their economies are constantly interacting with and influencing one another, we cannot map our economic future in isolation. Recognizing this, I cover the 'International Financial Architecture' in my next essay. When the financial crisis occurred in East Asia after July 1997, it generated enormous international concerns over financial stability in the world. This crisis was sudden and unanticipated. Neither the countries themselves nor the credit rating agencies or the international financial institutions anticipated a crisis of such magnitude. The speed with which the crisis was transmitted to other countries took the global financial community by surprise, and the burden of market failure was shifted to governments and the public sector. In the end, the international financial institutions, in particular the International Monetary Fund (IMF), came to the rescue of the affected countries, particularly Indonesia, South Korea and Thailand, which were the worst hit.

While still on the topic of the Asian crisis, one of the major factors relevant to it was the appropriateness of the exchange rate regime. We discuss this aspect in detail in the next chapter on 'Exchange Rate Management'. On exchange rate management, particularly for emerging markets, there is a consensus on the so-called 'impossible trinity', namely full capital account convertibility (CAC), monetary independence (for inflation control) and a stable currency. A study by the IMF shows that by far, the most common exchange rate regime adopted by countries, including industrial countries, is neither currency board nor a free float. Most countries have adopted various types of intermediate regimes including fixed pegs, crawling pegs, fixed rates within bands and independent float with periodic foreign exchange intervention. India has adopted the policy of flexible and competitive exchange rate, which is closely managed by the Reserve Bank of India (RBI). India's forex markets (short for foreign exchange markets) are relatively thin, and the declared policy of the RBI is to meet temporary demand–supply imbalances that arise from time to time. The objective is to keep market movements orderly and ensure that there is no liquidity problem or panic-induced volatility. On the whole, over time, India's exchange rate policy has been quite successful in the international context.

After the onset of the East Asian crisis in mid-1997, there has been a change in the perception about the role of financial system in the overall development of the country. Earlier, the real economy was supposed to lead and shape the financial system. During the Asian crisis, this perception changed dramatically. It became clear that it was the weak financial economy that had led to the collapse of the real economy. As such, proper development of the financial system was no longer regarded as an 'ancillary' or as adjunct to the development of the real sector, but as a necessary precondition of growth. The essay on 'Finance and Development: A Shifting Paradigm' deals at length with the lessons of the Asian crisis, India's experience regarding financial reforms introduced in the 1990s, and an appropriate agenda and priorities for the management of the financial system in future.

While the first section deals solely with the direct economic metrics, the second section goes beyond all such metrics to study other aspects which have a bearing on the economic growth of any country. Political reforms and governance play a critical role in how well an economy progresses and how well the economic progress gets translated into social equalities.

We begin this section with a chapter on the 'Role of Parliament'. The parliament is the supreme forum of India's democracy and represents the will of the people and their diverse identities.

Over time, there has been a subtle change in the role of the parliament in view of frequent disruptions that occur in the two Houses—Lok Sabha and Rajya Sabha. In the annals of India's parliamentary history, the events that took place during five days, between 18 March and 22 March 2006, were unique. A number of unexpected decisions were announced by the government regarding the business agenda of the two Houses, which were passively accepted by both the Houses. These decisions involved a major change in the established procedure for consideration of the Budget, a drastic revision of the business of the two Houses and a sudden adjournment of the parliament *sine die*, which was followed by a reversal of this decision a few days later. This chapter provides an eye-witness account of parliament in detail during these five days, which had never happened before in its history. There have also been several occasions when silences of parliament in respect of important issues put before the House have been just as loud as the prolonged debates on foreign policy, employment and development policy!

The next chapter deals with the role of the two other branches of the State, i.e., the 'Executive and the Judiciary'. A great deal has been written on the atrophy, non-accountability, corruption and inaptitude of the Indian administrative system. There is almost complete unanimity that despite having some of the best and brightest persons in the civil services, the system as a whole has become non-functional, and that there is very little possibility of reforming it. The worst sufferers of the weakness in the administrative system are the poor because of their dependence on public services and government programmes for various facilities, such as subsidized food and health services. This is the principal reason for the increasing disparities between urban and rural areas as well as widening in the income levels of different classes of citizens.

On balance, the long-term interests of the public and ordinary citizens are safer when the judiciary continues to be the watchdog of India's democratic conventions and the final arbiter of the Constitutional validity of the law or action approved by the legislature or the government of the day. The great advantage of the judicial review of the decisions taken by the executive and the legislature is that everyone, irrespective of their belief, has access to the courts. This is not the case in respect of the executive, parliament or state legislatures. Another advantage of the judiciary for the people is that, even if a judicial verdict is wrong or socially unacceptable, it is subject to review and reversal. This is usually not the case with legislative or executive decisions unless the government of the day so decides.

While a country may be progressing in economic terms, what is important is that the economic metrics reflect in social progress as well. If only the rich are getting richer, then that is not real progress. Poverty eradication is a critical goal for any developing country. What we see in India is that this goal has largely remained unrealized. The essay on 'Crisis of Governance' highlights the fact that while budget expenditure on anti-poverty programmes is relatively small, what is more disturbing is that as much as 70–80 per cent of the expenditure is on account of salaries of government servants at various stages of implementation of the programmes (from the central ministries to the village level). Over time, the process and procedures for conducting business in government and public-sector organizations, have also become largely non-functional. The essay emphasizes that to improve services to the public sector within a democratic framework, it is essential to impose greater accountability on both ministers and civil servants for properly implementing the policies announced for the benefit of the people. If transmission and distribution losses are reduced by even half through better management of the available budgetary resources, the improvement in the supply of services to the people, as well as financial savings, would be immense. It is also desirable to revert to a rule-based system of administration, which circumscribes the powers of politicians and confers greater authority to the civil service itself for self-regulation. The next essay on the 'Economics of Non-Performance' further elaborates on some of these issues.

It is extremely important to set in place checks and measures that ensure accountability. In the chapter on 'A Definitive Agenda for Political Reforms', I outline certain priorities to make the present political system more accountable and to strengthen the democratic process by improving the functioning of different agencies of the state, particularly parliament and the government, to ensure the speedy implementation of public policies. Some of the issues that need to be resolved as early as possible include review of the present administrative powers between the Union and the states; reform of the present system of election to the Rajya Sabha; introduction of a practical and equitable scheme for the state funding of elections; simplifying the present complex administrative procedures; reform of the legal system to reduce judicial delays in settling important cases; and separation of powers within the executive between ministers and civil servants in respect of some administrative matters, such as postings and promotions within a ministry.

What can we do to ensure that we are on the right track when it comes to economic prosperity for the country and for its people, without exclusion? 'India's Economy in the Twenty-first Century' (in Section I) deals with some economic measures which can be implemented to improve our country's prospects in terms of growth and poverty alleviation. Fortunately, there are very few developing countries that are as well placed in this century as India to take advantage of the phenomenal changes that have occurred in production technologies, international trade and deployment of skilled manpower. The shift in India's comparative advantage has substantial implications for its growth potential in the long term. While the COVID-19 pandemic of 2020 had derailed us—as indeed several other economies across the world—it is my belief that the pandemic will lose its sting in a year or less, and it should then be possible for India to achieve a growth rate of 7–8 per cent per annum, and virtually alleviate poverty over the next 25 years. India's competitive score, as revealed in the 'Global Competitiveness Index for 2017–18', remains high and India ranked 40 among 137 economies included in the index[2]. India's rank is one of the highest among all developing countries.

◆

[2]Chanchal Chauhan, 'India Slips One Rank in Global Competitiveness, Ranked 40th in 137 Economies', India.com, 27 September 2017, https://www.india.com/news/india/india-slips-one-rank-in-global-competitiveness-ranked-40th-in-137-economies-2500238/, accessed 16 March 2021.

# SECTION I
## LEARNING FROM OUR ECONOMIC PAST

# 1

## TOWARDS A BALANCE OF PAYMENTS VIABILITY

### 1991

Since the beginning of the Second Five-Year Plan (1956–61), India experienced BOP problems of varying intensity in 29 out of 35 years. Yet, a viable strategy to tackle this problem had eluded India's planners. During the same period, a number of developing countries, which started more or less with the same degree of industrialization and income level as India, had achieved remarkable success in increasing the share of manufacturing in total output and generating surpluses in their BOP. The cost for India had been heavy. Periodic crises had upset the planning process and reduced the room for manoeuvrability in fashioning macroeconomic policies in response to the changing domestic and international environment. These crises had also increased the country's dependence on external capital markets and also its vulnerability to external shocks.

### A Review of the Balance of Payments: 1956-57 to 1989-91

The period from 1956–57 to 1989–91 can be divided into three subperiods, depending on the nature of the BOP problem, the overall macroeconomic environment and the external aid situation.

These three subperiods are: 1956–57 to 1975–76 (Period I); 1976–77 to 1979–80 (Period II) and 1980–81 to 1990–91 (Period III). Period I and III were characterized by persistent BOP problems. However, there were important differences in the macroeconomic environment, particularly in the fiscal situation. Period I was also characterized by large inflows of aid on concessional terms, while Period III saw a large recourse to commercial loans and a substantial increase in external debt. During Period II, which was relatively brief, there was a substantial improvement in the BOP and foreign exchange reserve position. Let us look at these three subperiods more closely.

## Period I: 1956–57 to 1975–76

The entire period was very difficult for India's BOP, partly because of slow export growth in relation to import requirements, and partly because of adverse external factors. This period was marked by three wars (with China in 1962 and with Pakistan in 1965 and 1971); several droughts (including the disastrous droughts of 1965–66 and 1966–67); and the first oil shock in 1973. During this period, despite tight import controls (through quantitative restrictions) and foreign exchange regulations, the current account deficit (CAD) was 1.8 per cent of the gross domestic product (GDP). Foreign exchange reserves were low, less than necessary to cover three months' imports. Almost the entire CAD (92 per cent) was financed by inflows of external assistance on highly concessional terms. There was hardly any commercial debt.

A striking feature of this period was fiscal conservatism. The fiscal deficit (of the Centre, states and union territories) constituted less than 6 per cent of the GDP (as against 10.5 per cent of the GDP in 1989–90). The conservative fiscal policy also had implications for monetary policy. The average growth rate of money supply was only 11 per cent (as compared to 19.4 per cent in 1989–90). The rate of inflation was also relatively low

(6.7 per cent). A notable feature was the increase in public savings from 1.9 per cent in 1956–57 to 4.2 per cent in 1975–76. The average annual GDP growth rate was, however, only 3.9 per cent during this period. There was also a significant increase in the capital–output ratio, particularly because of low productivity and partly because of a shift in the structure of investment to more capital-intensive sectors.

## Period II: 1976-77 to 1979-80

This phase saw the golden years for India's BOP. There was a small current account surplus (0.6 per cent of the GDP on an average) and foreign exchange reserves equivalent to about seven months' imports. The export growth was good, but the primary reason for the sharp improvement in BOP was a dramatic improvement in net invisibles (mainly on account of remittances). Net invisibles increased from a paltry ₹193 crore in 1974–75 to ₹2,486 crore in 1979–80. The increase in foreign assets led to a substantial acceleration in the growth of money supply during this period (about 20 per cent per annum). However, the rate of inflation was surprisingly low. Excluding the drought year of 1979–80, the rate of inflation was barely 5 per cent per annum. Fiscal deficits, as a percentage of the GDP, were also moderate (about 7 per cent per annum on an average).

This period also saw the beginnings of a slow process of industrial and import liberalization. The import policy was gradually relaxed and a greater 'automaticity' was introduced in the policy relating to the imports of raw materials and components for domestic production. The most important change in import policy was introduced in 1978–79, when all items not specifically restricted or banned were listed under an Open General Licence (OGL) category, meaning these could be freely imported for domestic production. There were also some changes in the industrial licencing system, which facilitated an expansion of output in important sectors (such as cement). The exemption limit for the requirement of an industrial licence was raised from ₹1 crore (of investment in plant and machinery) to ₹3 crore.

The BOP situation, however, changed significantly in 1979–80 because of the second oil shock and the deterioration in India's terms of trade. The full impact of the increase in oil prices was reflected in the trade balance of 1980–81. The trade deficit increased from ₹2,200 crore in 1978–79 to ₹3,400 crore in 1979–80, and further to ₹6,200 crore in 1980–81. The impact of the increase in the trade deficit on reserves was also moderate, because of the increase in remittances and net invisibles in 1979–80.

## Period III: 1980–81 to 1990–91

In 1981, India entered into an arrangement with the IMF for a loan of special drawing rights (SDR) 5 billion under the Extended Fund Facility. The amount was to be disbursed over a three-year period. India, however, drew only SDR 3.9 billion and the arrangement was terminated in early 1984, at India's request, because of the improvement in the BOP position in 1983–84. The IMF loan helped to maintain reserves at a comfortable level in the first half of the 1980s, even though the CAD had increased after 1979–80. There was also a sharp increase in the indigenous production of oil, which reduced the growth in total imports.

## Balance of Payments Policy: Future Directions

India's Five-Year Plans were consistent in underestimating the CADs implicit in the plan projections of investment and income growth. Except for the Fifth Plan, import requirements were consistently underestimated. This may have been partly due to underestimation of growth in maintenance imports and partly due to over-optimistic assumptions regarding the possibility of import substitution. Industries which were expected to substitute imports, in fact, registered large increases in their own import requirements of raw materials and components (e.g. petrochemicals, automobiles and electronics). The underestimation of the CAD resulted in frequent BOP crises, the under-funding of projects, delays and cost escalation.

It is also fairly well established that widening fiscal deficits were a primary cause of BOP problems. An important priority in future is to reduce the government's fiscal deficit. In addition, certain modifications in the import–export policy and the policy for commercial borrowings are necessary.

## Fiscal Deficits

A major difference in the BOP crises of the 1960s and '80s was the macroeconomic environment prevailing in the country, particularly the deteriorating fiscal situation. The rising fiscal deficits in the 1980s were adversely affected by the BOP situation in two ways. First, the fiscal deficit and the increasing recourse of the government to the banking sector and the RBI for financing these deficits was the primary cause of the expansion in the aggregate money supply in the economy. The expansion of aggregate demand resulted in higher inflation and a higher import demand. Second, the increase in fiscal deficits led to a greater reliance on external commercial borrowings to finance public-sector investments. About 90 per cent of the approvals for external commercial borrowings (in value) were given to meet public-sector requirements during 1985–86 to 1989–90. While the bulk of the external debt was a liability of the public sector (and the government), the public sector's contribution to exports was marginal. The large external commercial borrowings of the past added to the debt-servicing burden and rendered the task of BOP management more difficult.

In its quest for BOP viability in the 1990s, it was crucial for India to give first priority to measures for reducing the fiscal deficit. However, the fiscal package had to be supplemented with other measures.

## Export-Import Policy

Within the broad framework of the existing export–import policy of the time, there were a number of changes which could have been introduced to tilt the balance in favour of exports as against production for the domestic market alone. It was *not* necessary to completely dismantle the prevailing trade regime to achieve a rapid growth in exports. Many administrative rules could have been turned around to encourage new exports, just as they were used earlier to promote new import-replacing activities. During a period when the availability of external finance is extremely limited, the least-risk strategy is *first* to get exports up and then to think of liberalizing imports. Import liberalization without adequate export growth or an abundant supply of external finance can be self-defeating.

The first task of reform in export–import policy should have been to shift the balance of profitability in favour of exporters. In order to achieve this objective, it was not necessary to destroy the import-substituting industry, but to shift its focus to the export market. Many import-substitution industries developed during the preceding 40 years were not uncompetitive and technologically backward. With a proper incentive structure, which rewarded exports adequately, the existing productive capacity could be made to contribute to export expansion. The main reason for the relatively low profitability of exports, as compared to domestic sales, was the high level of tariff and non-tariff protection granted to import-substituting industries. This raised domestic prices substantially above export prices. The existence of budgetary constraints made it easier to tax imports than to subsidize exports. As budgetary constraints were likely to continue in the future, the best instrument for increasing the profitability of exports was to shift the additional burden of export subsidization to importers. This could be done by a more aggressive use of the replenishment (REP) licence mechanism for financing imports.

The prevailing import control system permitted imports of raw materials and components by domestic manufacturers, irrespective of whether sales were at home or abroad. These were generally referred to as imports by 'Actual Users' (AU). Such imports could be under the OGL or under the restricted list. The former did not require prior clearance or a licence. Items on the restricted list were, however, subject to administrative scrutiny and quantitative restrictions, and could be imported only against a licence. The principal criterion for the classification of items into different licencing categories was indigenous availability. REP licences provide an additional import facility to exporters, which was linked to the value of exports. Imports under REP and other export-related schemes did not affect the entitlement of the manufacturer for imports under the AU scheme. REP licences could be used to import a wider range of commodities (including banned items, to a certain extent). They were also freely transferable. On transfer, REP licences generally enjoyed a 'premium' in the market, which provided some additional subsidy to the exporter.

The subsidy available to the exporter from the sale of REP licences varied according to the restrictiveness of the AU policy and the size of the REP market. The subsidy was paid by the importer to the exporter, who forwent his right to import. It followed that if all imports were to be permitted only against flexible REP licences, an explicit link would be established between exports and imports. Exporters would earn a higher subsidy and gain, and the costs of sale of goods in the domestic market would go up, depending on their import content. *Ceteris paribus,* export profitability would go up while domestic profitability would decline.

Another important area of reform at the point was the duty-drawback system. A simple and liberal duty-drawback system had proved highly effective as an export-promotion measure in several countries. What was required was that the duty paid on capital goods imports by exporters be brought within the ambit of the drawback system by enhancing their rates across the board. The indirect exporters, i.e., those who supply domestically manufactured inputs to exporters, also needed to be covered by the drawback system. This could be done through the introduction of domestic letters of credit, which had been tried in other countries. The duty-drawback rate structure needed to be simplified, such that it was not based on a firm-by-firm scrutiny of duty paid.

There was also an urgent need for an administrative simplification of rules and regulations, and for making the system more stable. The case-by-case administrative scrutiny of imports/ exports deserved to be totally eliminated over a period of time. On the import side, all restricted items were needed to be shifted to REP licences or OGL with higher tariffs over the next two or three years. On the export side, it was important that there were no quantitative or quota restrictions on exports. In many cases, these restrictions were redundant as domestic prices were substantially higher than international prices. The one area where they were effective was agricultural exports. Quota restrictions on agricultural exports had the effect of reducing the total income of farmers in cases where international prices were higher than domestic prices. This was not equitable. Another important administrative measure was to move towards uniformity of rates from a system of highly differentiated rate schedules. This would make the system more transparent and less amenable to sectional pressures.

A realistic exchange rate is obviously crucial to export performance. There is some evidence that in earlier years, this instrument was not used sufficiently to support India's exports. Fortunately, the management of the exchange rate has ceased to be a matter of controversy in our country in recent years. While there may be a difference of views among experts regarding the exact point at which the rupee exchange rate should be placed vis-à-vis other currencies, the area of difference is not very large.

## Invisibles

The net receipts from invisibles provided valuable support to India's BOP for nearly a decade after 1975–76. After 1985–86, although there was some growth in private transfers, it was not sufficient to offset the increase in interest payments on external debt. In 1989–90, the contribution of invisibles to financing the trade deficit declined to about 20 per cent, as against nearly 50 per cent in 1984–85. Among net invisible receipts, two main sources were foreign travel and private remittances. Net receipts from foreign travel also increased substantially in the 1980s, and accounted for about 30 per cent of the total net invisible receipts towards the end of the decade.

In the 1980s, non-residents in the Middle East and Gulf countries were estimated to account for roughly 45 per cent of total private remittances to India. The future flow of remittances (as against non-resident deposits) was considered to be heavily dependent on the growth of the economies of the Middle East after the recent Gulf War, and their policy towards Indian immigrants. In the early 1990s, there were considerable uncertainties on both counts.

There was very little empirical research on macroeconomic factors affecting remittances into India. In the absence of such research, it was difficult to come to any firm conclusions on the policies to be adopted by India to encourage the flow of such remittances into the country. Obviously, the exchange rate is an important factor, particularly the differential between the 'official' exchange rate and the 'unofficial' exchange rate. If the divergence is too great (in favour of the unofficial rate), then flows of remittances through the banking channels are likely to be affected adversely. In such an event, there is a strong case for narrowing the gap between the two rates by depreciating the official rate. In India, at that point, there was no evidence, although data was very poor, that the divergence was too large or had been consistently widening. The differentials had remained fairly stable and the unofficial rate had moved in line with the official rate. In a regime of import-and-exchange controls, the elimination of the differential was not feasible, as unsatisfied demand for foreign exchange could spill over to the unofficial market, whatever the level of the official exchange rate.

In respect of remittances, the most effective policy may have been to follow a realistic and competitive exchange rate policy, and to provide efficient and cheap banking services. Gifts in foreign exchange already enjoyed tax exemption. This fiscal concession should be extended to other forms of receipts in foreign exchange by residents.

## Capital Account

The structure of capital account in India had undergone considerable changes in the three decades preceding the 1990s. In the 1960s and '70s, the main sources of financing the CAD were official loans on concessional terms. In 1980, nearly 90 per cent of India's outstanding debt was to official creditors and 85 per cent of it was on concessional terms. By 1989, however, the picture had changed dramatically and the official loans accounted for 61 per cent of the total outstanding debt, and the proportion of concessional loans had fallen to 47 per cent. Incidentally, almost the whole of the decline in concessional assistance was due to a change in the composition of the World Bank Group lending to India. During the 1980s, our country was the poorest among the developing nations, and on objective grounds, India's claim for concessional assistance was unquestionable. Nevertheless, the World Bank Group reduced flows of concessional assistance to India.

In India, issues of relevance from the BOP point of view primarily related to foreign direct investment (FDI) and commercial borrowings. Official multilateral and bilateral flows were, of course, also important. However, with these flows being mainly supply-determined, India could hardly do much to affect their volumes. The volume of FDI in India was very small. It was of the order of ₹200 crore (or roughly $100 million) on an average of three years (compared to $3.4 billion in China, in 1989). India followed a restrictive policy with regard to foreign investment on the grounds of promoting the growth of indigenous capital. The main restrictions related to the percentage of equity that foreigners could hold and the areas in which foreign investment could take place.

In India, there was a strong case for simplifying the laws relating to foreign investment and for eliminating the bias against equity investment. The volumes at that point were so small that even a manyfold increase could do no harm. With the globalization of production in the world economy and the integration of capital markets, the case in favour of equity investment had become stronger. The only thing which needed to be ensured was that such investment was positive in terms of foreign exchange and was internationally competitive. In respect of competitive investments, the rules governing entry were needed to be made automatic and transparent. However, it was likely that new equity investment would flow into industrial projects for setting up new capacity. The BOP difficulty existing at that point in time was one of financing maintenance imports and debt-service obligations. It was realistic to assume that higher equity investment by itself would not provide an adequate source of financing.

On the whole, it was felt that, in future, India must find the means to move to a more viable BOP situation. Regular and persistent BOP problems, as witnessed in 35 years, up to 1991, were avoidable. There was sufficient international experience available to guide policymakers towards devising a new strategy to deal with the problem and for freeing development planning from the tyranny of avoidable payments crises. Fortunately, the magnitude of the BOP correction required in India was relatively small by international standards—no more than 1–1.5 per cent of the GDP. A big effort over the next two–three years, particularly in the fiscal field, was thus required. This effort, in any case, was necessary to remove another important constraint on India's development, i.e., the savings constraint. It was necessary also to tilt the balance of policies in favour of exports, as against domestic sales alone. This could be done without destroying the established industrial structure. The existing industrial capacity could be turned around to support export expansion with the introduction of suitable changes in the incentive structure. At that time, in view of external financing constraints, the least-risk strategy was first to get exports up and to then think of liberalizing imports, and not the other way round. The simplification of the tariff system and other measures to reduce administrative forms of intervention, including the abolition of the capacity licencing system, would make the task of export promotion easier. In view of the high debt/export ratio, there was no further scope for financing inward-looking, capital-intensive investment projects through external commercial borrowings.

# 2

# AFTER THE CRISIS:
# THE NEED FOR A NEW STRATEGY

## 1992

The seven-year period from 1984–85 to 1990–91 was marked by a high rate of industrial growth. The overall index of industrial production grew by 8.5 per cent per annum and the manufacturing sector grew by 9 per cent per annum during this period. A steady growth rate of this order over a fairly long period was achieved after a gap of two decades. Contrary to popular impression, industrial growth was fairly widespread and *not* concentrated only in certain sectors, such as automobiles, electronics and consumer durables. These sectors did register a very impressive growth, but their contribution to the overall growth rate of industrial production was relatively modest because they did not have a large weight in the index.

In 1991–92, however, in view of the BOP crises, the index of industrial production was practically stationary and there was no growth in the overall industrial production. There was a significant increase in the output of several infrastructure industries, such as electricity and coal, but this was more than offset by a decline in manufacturing production. Production of manufacturing industries fell by about 1.8 per cent during this period. Within the manufacturing sector, the sharpest decline occurred in the production of capital goods. This sector, which had shown high growth rates in the 1980s, registered a decline of 17 per cent in 1991–92. The other industrial group that showed a similar decline in production was consumer durables, in which output declined by about 14 per cent.

In 1990–91, in addition to the direct effects of an increase in oil prices on the BOP, India also faced a 'liquidity squeeze' because of the withdrawal of deposits by non-resident Indians (NRIs) and cessation of lending by commercial banks. As a result, import compression measures became severe, which naturally affected the growth of output in 1991–92 and besides, inventory levels were already low. In countries such as India, where import to GDP ratios are low, a 'forced' cut in imports tends to affect output more than proportionately.

After April 1991, the effects of import compression were further accentuated by a number of factors on the demand side. First, the collapse of the rupee trade with former Union of Soviet Socialist Republics (USSR) and the recession in Western countries affected the overall demand for exports in traditional markets. Second, the fiscal adjustment and slowdown in government expenditure, which was essential to stabilize the economy, affected demand for investment goods. Third, the increase in interest rates, which was necessitated by overall monetary considerations, increased the cost of investment and private demand for credit. All these factors, combined with continued import compression, had an adverse effect on output.

This is the background against which it is desirable to consider issues relating to the strategy for accelerated industrial growth in 1992–93 and beyond. This essay concentrates on investment, which is central to and crucial for industrial growth.

A study by Maurice Scott[1], published in 1992, after a careful analysis of data on the developments in the United

---

[1]Maurice Scott, 'Policy Implications for a New View of Economic Growth,' *The Economic Journal*, May 1992.

States (US) and the United Kingdom (UK), had concluded that 'investment' is a much more important proximate cause of growth than conventional theory. The large body of empirical growth-accounting studies show that changes in capital stock and employment often account for only about half of the output growth, the remainder being generally attributed to 'technical progress'. According to these studies, the increase in total factor productivity is more important in determining the growth rate of industrial output. Scott, however, argued that this result has depended crucially on defining capital and investment too narrowly—in terms of 'net' investment rather than 'gross' investment—and that if a correction is made for this factor, the rate of investment emerged as the most crucial variable for accelerating industrial growth. According to Scott, in addition to the direct effects of investment on growth, there are also large positive 'external' and 'learning' effects of investment, whereby investment by one firm creates and reveals investment opportunities for other firms.

Data for 19 developing countries, which have achieved a growth rate of 6 per cent per annum over a 25-year period (1965–1990), shows both a high growth rate of investment and a high level of investment during the entire period. Another study also found that in the 1960s and '70s, the relationship between the growth of income in developing countries and the growth of investment was even stronger than between the growth of income and the level of investment.[2]

In India's case also, there is some evidence that periods of high industrial growth were also periods of high growth in capital formation in manufacturing. Between Independence and the early 1990s, there have been three periods when our country achieved a growth rate of about 8 per cent per annum in industrial and manufacturing production. These periods were: 1954–55 to

---

[2]R.M. Sundrum, *Economic Growth in Theory and Practice*, Macmillan, London, 1990.

1965–66, 1975–76 to 1978–79 and 1984–85 to 1990–91. All these years also witnessed a substantially higher growth rate of capital formation in manufacturing. The 'causal link' between the two, or about which came first—growth or investment, is not clear. However, the association between the two is too strong to be ignored. It is most likely that investment and growth reinforce each other—higher investment leads to higher growth, which in turn, provides incentives for further investment.

One should not, of course, exaggerate the importance of increasing the 'quantity' of investment as against the 'efficiency' of investment. Efficiency and quality of investment are equally important, and a rise in the capital–output ratio (or a reduction in efficiency) can easily offset the gains from higher investment. This has happened in India in respect of several large projects. It is certainly possible to combine high investment with low growth. However, the converse is not true—there are very few cases of high industrial growth with low investment. It stands to reason that, without new investment in capital equipment and efficiency-enhancing technologies, industrial growth cannot be sustained for very long.

The basic point is simply that the key to accelerated growth in India, as elsewhere, lies in higher investment in industry. This is particularly so at low points in industrial and business cycles when the current output is growing at a slow rate. In looking for a strategy to achieve higher industrial growth, it is desirable to try and identify those factors which facilitate higher investment by industry in new plants and technology.

## Macroeconomic Stability

There is abundant evidence from past experience in India and other countries that macroeconomic stability is a necessary, though not a sufficient, condition for growth in industrial investment. Long-term confidence in the economy is weakened by instability, and investment is discouraged and rendered more inefficient. The desire to make quick gains from arbitrage and speculative activities overcomes the desire to make long-term investments in commodity-producing sectors. Instability can be the result of mistakes in macroeconomic policies for political or other reasons, or it may result from events over which the government has no control. Between 1988 and 1991, both sources of macroeconomic instability were present in India. It witnessed two general elections, three changes of government, several destabilizing domestic developments, the Gulf War (which imposed an extraordinary burden on the economy) and the collapse of the Soviet Union. The resulting macroeconomic instability, apart from causing direct financing and fiscal problems, inevitably affected the environment for long-term investments. The restoration of macroeconomic stability was, therefore, crucial for investment and growth. It was also an important objective of the stabilization programme launched by the government. If India succeeded in stabilizing the economy, despite some short-term costs, the medium- and long-term outlook for investment and employment would definitely improve.

## The Demand Situation

It has to be recognized that there may be a conflict between policies for stabilization and policies for boosting investment. Tight monetary and fiscal policies are good for stabilization, but they increase the risks associated with reduced investments. A situation of high profits and strong demand fosters growth, while low profits and weak demand are equally necessary to keep inflation in check. While in the run-up to the crucial decade of the 1990s, the growth of money supply in the economy had been high—close to 20 per cent in 1991—a number of industries were facing demand problems. This could have been due to a reduction in government expenditure, which is an important source of demand for many industries. It could also be due to a greater flow of funds into financial markets as well as the postponement of discretionary consumer expenditure by fixed-income groups in view of the price rise. The collapse of the Soviet trade and the recessionary conditions in the Organisation for Economic Cooperation and Development (OECD) countries would also have affected the demand situation for industries dependent on these markets. Whatever the proximate causes of a perceived slackness in demand, the fact remained that low profits and low demand were not good for investment.

Assuming that tight monetary and fiscal conditions were necessary for some more time to stabilize the economy, what could have been done to help investments? In the prevailing situation, it was unrealistic to expect any significant increase in public investment. Given the circumstances, the best strategy for industrial firms was to concentrate investment in areas not closely dependent on the existing domestic demand situation.

One obvious area was exports. The outlook for exports had never been more promising earlier. The export-profitability picture had improved substantially; policy hurdles in production and investment have been more or less eliminated and many firms had acquired a foothold (or at least a toehold) in foreign markets. Another potential advantage for India was that wage costs were rising much faster in several East Asian countries, which had shown dramatic export performance in the 1970s and '80s. India's share in world trade was less than 0.5 per cent, which meant that an expansion in its exports from new investments would not necessarily depend on an expansion of overall demand in the world economy.

Much of Indian industry suffered from obsolescence. In a more competitive environment in the future, only those firms were likely to survive which are constantly upgrading their technologies and their output mix. With the virtual abolition of licencing, new firms were likely to enter all profitable lines of manufacturing.

These firms were to have the advantage of having access to modern technologies and new products to meet existing demands. At the same time, existing firms had the advantage of location, cheaper real estate, a trained workforce and an existing marketing network. These initial advantages could become immensely valuable if new investments were made to replace old technologies and machines.

Another important area where there was immense potential for further investment, even under conditions of low final demand, was that of energy efficiency. Indian industry was largely an inefficient user of energy, as its energy use per unit of output was substantially higher than that in many other countries. It was reasonable to assume that with changes in our energy pricing policy and the exchange rate regime, over a period of time, the cost of energy in India would increase at a rate that is higher than the price of manufactured products. It followed that investment in energy efficiency, even under conditions of tight overall demand, could be an important source of higher profitability for industry.

These were some opportunities for investment not dependant on the state of the existing domestic demand. No doubt there were many more such potential areas. If a large number of units were to start investing in these and other areas, it would soon become a self-reinforcing process—with higher investment generating higher growth, higher demand and even higher investment.

### Finance

A well-functioning financial system is essential for the growth of investment. There are two aspects of the financing problem— availability and cost. So far as availability was concerned, more savings were likely to be available for private investment in future than had been the case in the recent past. The government was committed to reducing its fiscal deficit, which simply meant that its borrowings from banks and households were going to decline as a percentage of the country's GDP. Thus, domestic savings, which would have otherwise gone into government consumption or public investment, would be available for private investment. This was a crucial point. If the government borrowed less, there was to be more money available for investment elsewhere. For this to happen, however, it was necessary to ensure that a reduction in government borrowings did not lead to a fall in incomes and aggregate savings. This further underlined the need to create a favourable environment for investment, as growth and savings in the economy could be maintained only if investment grew. Otherwise, both savings and incomes would settle at a lower level.

There had been an increase in the cost of capital in the past few years, and the long-term real rates of interest (i.e., the nominal rates *minus* inflation) in the economy were about 6–7 per cent. These rates were higher than those prevailing in industrialized countries as well as in some semi-industrialized countries. In view of the scarcity of capital in India, one would expect real interest rates to be higher than those prevailing in capital-surplus countries, and most economists would probably agree that from a long-run point of view, there was a case for real interest rates to be lower than they are. There were, however, two problems in achieving this objective. First, India continued to have a highly complex structure of interest rates (despite recent improvements) with several highly concessional rates. As a result, normal or non-concessional interest rates tended to be higher than they should be. Second, the growth of money supply had been higher than targeted in recent years. Interest rate policy, therefore, had to respond to the need for restraining money supply.

It was to be hoped that the macroeconomic situation would stabilize sufficiently soon to permit further rationalization of the interest rate structure. There is, however, one change that could be introduced in interest rate policy without affecting the current stance of monetary and credit policy. During a period when interest rates are high for contracyclical reasons, there is a tendency on the part of entrepreneurs to postpone investment decisions to avoid 'locking-in' at prevailing high interest rates. Most countries had, therefore, introduced a system of variable interest rates, whereby interest rates on long-term investments varied every six months or so in line with movements in the market interest rates and the overall macroeconomic situation. A similar system of variable interest rates introduced in India for long-term funds would result in there being no incentive to postpone investments in anticipation of a fall in rates. A system of variable interest rates was also likely to be favourable for banks and financial institutions.

## Fiscal Policy

The role of fiscal policy in promoting investments in general, or in special regions, has long been debated by fiscal experts. By the early 1990s, a number of countries, including India, had experimented with various types of investment allowances or deposit schemes. While no conclusive evidence on the efficacy of these schemes in promoting aggregate investment was yet available, the prevailing consensus among fiscal experts was largely in favour of moderate corporate rates of taxation without special incentives rather than a system of high rates combined with special deductions for investments in new plants or particular regions. The former system was expected to be less distortional and avoided the possibility of a wasteful use of resources in low-yielding and inefficient activity merely to take advantage of fiscal privileges. There was, however, a strong case for depreciation provisions to be generous, particularly if the rate of inflation was relatively high. The replacement of capital has to take place at current prices, while depreciation is at the book value. During periods of relatively high inflation, liberal depreciation provisions are required to avoid taxation of 'illusory' profits.

Another investment-related issue in the area of fiscal policy is the taxation of capital goods imports. There was a conflict here too, as a well-established capital goods industry, because of various factors, needs tariff protection. At the same time, to reduce costs and make user industries more competitive, it was necessary to reduce import duties to more reasonable levels.

## Import Policy and Foreign Collaborations

Policies relating to imports, foreign collaborations and foreign investments had already been considerably liberalized by the early 1990s, and there seemed to be an extremely attractive environment for new investments. Problems that persisted should have been sorted out easily, as the government's overall approach was very clear as both import policy and foreign investment policy were supportive of domestic investments.

## Infrastructure

Policies for private investment in the infrastructure and power sectors had also been liberalized, and these would have provided an additional avenue for higher aggregate investment. However, at that point in time, the primary responsibility of providing infrastructure facilities and power remained with the government, both central and state. In the changed environment, where the government was reducing its role in licencing and controlling economic activities, the most important contribution of state governments would have been to improve the operational efficiency of infrastructure and power plants under their command. The old secretariat-oriented management style would not work, and a new management structure for state public-sector infrastructure industries had to be created as soon as possible.

## An 'Efficient' Public Sector

In 1988–89, as much as 53 per cent of the GDP in mining, manufacturing, construction, power and finance originated in the public sector. In 1960–61, the contribution of the public sector to the GDP in these areas was only 11 per cent, and in 1980–81, it was about 40 per cent.

With a contribution of more than half the GDP in manufacturing, mining, construction, and power and finance (taken together), it was obvious that the future of the economy would be closely linked to the performance of enterprises in these sectors for some time. It was not a question of profitability alone, but also of their contribution to production and growth of investment in the economy. The government was no longer in a position to subsidize their operational losses or to provide adequate resources for further investment. In this situation, their survival and growth were likely to depend increasingly on their own performance. However, it was important to remember that what happens to public-sector enterprises cannot be a matter of indifference to the rest of the economy. These enterprises are closely linked to other enterprises and sectors as suppliers as well as buyers of goods and services. If the public sector fails to function, so will the rest of the economy—at least over the next five to 10 years.

To conclude, on the whole, India's industrial future was highly positive. It was to be expected that India would be able to achieve an industrial growth rate of 8 to 9 per cent per annum not only in the 1990s but also in the long run. This order of growth was achieved in half the previous 40 years, despite various problems and several domestic and external setbacks. In the 1980s, a growth rate of this magnitude was achieved in eight out of 10 years. In the 1990s and beyond, India was expected to do even better. The key to higher industrial growth was efficient investment, and this is where India had to concentrate in the 1990s.

# 3

## TRADE, INVESTMENTS AND CAPITAL FLOWS

### 1996

If we look at the mid-1990s, India was among the least open economies in the world. Compared to East Asian or Latin American averages, India was virtually a closed economy, with a share of less than half a per cent in the world trade. There was nothing wrong with this except that the isolation in trade has been associated with persistent BOP problems requiring emergency assistance from abroad or periodic borrowings from the IMF. According to one calculation, during the period of 35 years between 1956 and 1991, India experienced BOP problems of varying intensity in as many as 29 years. Thus, the post-Independence, inward-looking strategy of development, which was supposed to make India economically strong and self-reliant, turned out to be one which, after a few initial years, made it increasingly dependent on periodic international rescue operations.

### Effects of Trade Isolation

It was necessary to keep this background in view while considering the trade policy options for India in the future that lay beyond the 1990s. It was simply not true that low volume of imports and exports made India 'self-reliant' and less dependent on the external world. It only changed the form of dependence. Nor did India prosper industrially. Our industrial performance was quite poor by international standards. The growth rate of production was low and technological progress was tardy. The missed opportunities in international trade in manufacture did not generate employment. The growth of real wages of industrial labour in India was also much lower than those in countries where industry and trade grew faster. There is no doubt that if India had maintained its share of world trade at the same level as in 1950 (i.e., 2 per cent), both the income and employment situation would have been much better than they were four decades later.

India's planning process was preoccupied with the question of 'priorities', and one reason for trade isolation was to ensure that industrial development took place along the lines of Plan priorities. For example, under the Plan, the production of capital goods and the development of engineering industries were encouraged, while the production of consumer goods was discouraged. Thus, imports of capital goods and intermediates were permitted only on grounds of essentiality, while those of final or consumer goods were banned or severely restricted. The priority-based import policy, however, gave the wrong signals to industry as domestic production of banned and restricted goods became more profitable than those of permissible items. India, for example, became self-sufficient in automobiles (a non-priority banned item in the Plan), but dependent on imports of fertilizers (a priority item). This experience does not support the hypothesis that a low volume of trade was conducive to the allocation of national savings according to national priorities.

India's loss of markets abroad was mostly policy-induced. This is exemplified by the case of textiles, India's largest industry. Our country had enjoyed a comparative advantage in this industry for long. Yet, within a few years, India whittled away its traditional advantage by severely restricting the output of cloth (by imposing licencing restrictions), increasing costs (through widespread interventions in the labour and cotton markets) and preventing induction of innovative technology and new fibres. In 1973–74, India's share in world markets for textiles and clothing exports was the same as that of China (4.5 per cent). Korea's share was 7.7 per cent and that of Pakistan was only 1.7 per cent. By 1985–86, China's share increased to as much as 14.6 per cent and that of Korea to 13.6 per cent. Other developing countries such as Pakistan, Thailand and Hong Kong also increased their shares substantially. India was the only country among major developing country/exporters whose share actually fell from 4.5 per cent to 3.8 per cent during this period (1985–86).

Three distinct phases are discernible in India's approach and policy towards exports. The early phase, which lasted up to around 1972–73, was one of extreme export pessimism. It was believed that the terms of trade of developing countries were destined to deteriorate over time regardless of the policies of developing countries. This was a crucial assumption as it firmly established a case for discouragement of exports and for policies that encouraged production for the domestic market. As it happened, however, the 'export pessimism' thesis was not borne out by post-War developments in international trade. Trade in the 1950s and '60s grew faster than world income and several developing countries showed sharp increases in their share of world trade. Exports of manufacturers from developing countries grew twice as fast as the industrial countries' income in the 1960s and four times as rapidly in the 1970s.

By the late 1960s, the domestic consequences of India's trade isolation and loss of competitiveness were also becoming apparent. Domestic industrial production was stagnating, costs were rising, employment growth was slow and industrial sickness was spreading. Unfortunately, an overhaul of trade policy was not possible as India, by then, was trapped in a low level equilibrium characterized by low growth, periodic external and domestic shocks (e.g., droughts and wars) and persistent BOP problems with very little room for manoeuvre. By then, a powerful domestic industrial lobby had also emerged. Among domestic industrial interests which were opposed to trade liberalization, there were hundreds and thousands of small-scale units in all parts of the country. These provided substantial employment and enjoyed total protection from competition from large domestic units or imports. Thus, both practical considerations (e.g., shortage of foreign exchange) and political economy compulsions (e.g., labour and industrial interests) committed India to the continued pursuit of inward-looking, anti-trade policies.

In 1973, India was confronted with one of the most severe BOP crises due to the sharp rise in international oil prices. In addition to imports of oil, India had also become dependent on large imports of fertilizers, steel, non-ferrous metals and capital goods. While import requirements were increasing faster than visualized, exports were stagnating. A re-examination of India's export strategy could no longer be avoided. International assistance also became increasingly conditional on improvement in export performance. In the mid-1970s, a number of changes were made in industrial and trade policies, which provided special treatment to exports. Thus, for example, it was provided that industrial capacity used for exports was not to be counted as part of licenced capacity. Simplified procedures were also introduced for imports of technology, raw materials and capital goods for exports. Export processing zones were established and promotional agencies set up to promote exports. Financial incentives for exports were enhanced and the exchange rate policy was used in support of exports. These policies seemed to work for a while and export growth was quite robust in the second half of the 1970s. However, this trend did not last long.

In this middle phase, which lasted until the end of the 1980s, exports were still regarded as something 'exceptional' and not as an integral part of domestic industrial policy. The basic framework of industrial and trade policies remained virtually unchanged. A fundamental transformation of the trade regime was not effected until after the economic crisis of 1990–91. By then, it was clear that India's closed trade regime was no longer sustainable. India's fiscal situation had also deteriorated and there was no scope for providing further direct or indirect budgetary subsidies for exports. Under these circumstances, sustained growth in exports over the long run was not feasible without a change in the incentive framework and elimination of anti-export bias in trade policies.

Beginning 1992, the government effected certain major changes in import and export policies. Exchange restrictions on current transactions and various types of quotas for import of goods, with the exception of most consumer goods, were abolished. The items subject to export restrictions were drastically reduced. The exchange rate, after two substantial devaluations, was largely determined by the market (although the RBI intervened in the market from time to time). Import duties, which were among the highest in the world, were brought down progressively. Maximum tariffs were reduced from over 400 per cent in 1990–91 to 50 per cent in 1995–56, with some exceptions. The average tariff (weighted by volume of imports were reduced from 87 per cent to 27 per cent).

These were important changes and have helped to reduce the anti-export bias in India's international trade. The effective exchange rates for exports (inclusive of duty drawbacks and other export benefits) were closer to that for imports (inclusive of all duties). This narrowed the gap in profitability of sales in exports and domestic markets. The results were favourable. In 1993–94 and 1994–95, exports increased by nearly 20 per cent per annum, and financed 90 per cent of imports (compared to 60 per cent in the latter half of the 1980s). Domestic industry experienced a broad-based recovery with industrial growth of more than 10 per cent in 1995–96 (as against 0.5 per cent in 1991–92). The capital goods sector, despite import liberalization, grew by nearly 25 per cent in 1994–95. This experience was contrary to the popular belief that openness in trade is injurious to the health of domestic industry.

Certain broad conclusions can be derived from the above discussion. First and foremost, the 'export pessimism' of the 1950s and '60s cost India dearly. India cannot afford to make the same mistake again. India must, therefore, firmly dismiss the arguments in support of a new wave of 'export pessimism' (such as protectionism in industrial countries) or 'economic nationalism' (based on the belief that domestic production for domestic consumption is economically superior to trade). Rise in protectionism abroad must be resisted. Till the mid-1990s, India's share of the world trade was only 0.4 per cent and so, there was plenty of scope for India to expand its share even if growth of world trade slowed down. Similarly, there is no virtue in production being 'domestic' if such production is inefficient, and a country becomes more dependent on emergency assistance from abroad to meet its essential import requirements.

Second, export policies have a lot to do with export success. Supportive fiscal and exchange rate policies are clearly important. There is the need to use both 'generalized' policies which improve efficiency in the use of resources in the economy as a whole and 'special' policies which provide preferential support to exports (e.g., in respect of access to credit). Economists are used to thinking in terms of one set of 'right' policies to the exclusion of others. The real world is much more complex and success may require the pursuit of a mix of policies which do not fit neatly into any fixed paradigm.

Third, there is some evidence that in the 1970s and '80s, periods of high growth in the manufacturing sector were also periods of high growth in exports. The two are likely to go hand in hand, provided the overall macroeconomic policies are supportive of exports.

## Trade Policy Options

The above review of past policies and experience is necessary as, in view of India's colonial past, a discussion of trade India issues arouses strong passions in the country. During British rule, trade was indeed an instrument of economic exploitation and there was substantial drain of resources from India to Britain. 'Free Trade' was used as a mechanism to transfer raw materials cheaply to Britain, which in turn, processed them into final goods for export back to the colonies. The inequity inherent in this type of trade figured prominently in the national debate at that time, and it was neither surprising nor unreasonable that, after Independence, India was highly suspicious of trade as a positive instrument of development policy. However, the fact remains that anti-trade policies were continued for too long and were carried to extremes even after the conditions of international trade had changed fundamentally.

An important objective of India's trade policy in the future must be to recapture the lost ground in India's share in world trade. India's share declined from 2 per cent in 1950 to 1 per cent in 1960 and to less than half a per cent by the mid-1970s. A feasible goal during the mid-1990s was to raise India's share of trade to at least 1.5 per cent in the next 10 years, that is, by about 0.1 per cent per annum. This would require that India's exports and imports should increase substantially at a faster rate than the growth of GDP or that of international trade in general. Both exports and imports had to grow in a balanced manner so that there is no recurrence of a BOP problem or a substantial addition to commercial debt.

As mentioned earlier, exports had done well in the couple of years preceding this time. However, there was still a long way to go. While the anti-export bias in trade policies had been reduced through the lowering of tariff rates on imports (thereby reducing profitability of domestic sales), import duties were still high by international standards. The tariff rate in most countries, including developing countries, was generally less than 10 per cent on industrial imports.

Tariffs in India on such imports were four or five times as high. In principle, further reductions in tariff rates to international levels were desirable in order to remove the remaining bias against production for exports. Fortunately, India's reserves were strong and the exchange rate was highly competitive. A further reduction in tariff levels, therefore, could take place without endangering the BOP or leading to a fall in domestic output. At the exchange rate of nearly ₹35 to a dollar (in December 1995), domestic output was unlikely to be displaced by imports except in cases where value added, at world prices, was very low or negative.

India's domestic interest rates and inflation rates were substantially higher than those prevailing in industrialized countries and in East Asia. This posed a problem for exports, as high interest rates increase the cost of production and lower the expected returns from investment. This is a thorny issue for monetary policy because cross-subsidization, through fiscal means, is neither desirable nor feasible in view of budgetary constraints. In this situation, a practical remedy is to permit exporters to borrow abroad at international interest rates. In any case, established exporters should have the first claim and access to international commercial borrowings for investments as well as bank credit for production (depending on the share of exports in total output). A further measure to insulate exporters from high domestic interest rates is to create special domestic financing facilities in term lending institutions for export-related investments. These are no doubt clumsy and partial answers. A fundamental solution lies in reducing domestic inflation and interest rates to international levels. This, however, was not likely to happen until fiscal and public-sectors deficits were reduced to more reasonable proportions.

Infrastructure for exports also required special attention, and should have been a priority area for investment. In the mid-

1990s, the state of essential facilities (such as ports, airports, communications and transport) were well below international standards. The clearance and processing facilities were time-consuming and costly. They could no longer cope with the demands of fast-changing technologies and new patterns of international trade. In order to upgrade existing facilities and create new ones, it was necessary to induct market-oriented business techniques and to introduce greater competition in respect of ports, communication and other facilities. These would also benefit the labour employed in the provision of these services. Greater competition was likely to create new opportunities for employment and lead to a rise in wages.

In the past, India had intervened heavily to promote exports, but the 'quality' of such intervention had been poor as compared to policies pursued by successful exporters like Japan, Korea, Malaysia and China. Unlike India, their intervention was strategic and not occasional. They encouraged competition in the domestic market; India fragmented its markets. India concentrated on identifying products and providing specific amounts of assistance on a case-by-case basis. The Japanese or the Korean system, on the other hand, was concerned with altering the generalized incentive framework in favour of exporters, rather than with specific products and specific markets. What was required was that India intervened aggressively to promote exports (as did most other countries, including industrial countries like Japan and the US). Such intervention needed to be strategic and directed at improving the domestic environment for exports and opening markets abroad.

Even in the mid-1990s, the export policy for agricultural commodities continued to be restrictive. A case had been made in the trade policy literature for liberalizing agricultural exports. The main argument in favour of liberalization was that farmers in effect enjoy 'negative' protection since world prices of several commodities are higher than those in India. Export controls prevent farmers from enjoying higher returns, and thereby discourage growth in productivity and capital investment. In principle, this argument was valid and there is a strong case for liberalization of agricultural trade, particularly in respect of crops like cotton and sugar. However, in view of the importance of the agricultural sector in terms of GDP, employment, poverty levels and consumer welfare, a change in the policy framework had to be effected after carefully considering the 'risk' factors. Any change that destabilized established production patterns or generated adverse expectations could create considerable economic difficulties and misery for millions of people.

Thus, for example, a sudden freeing of exports of foodgrains could lead to a sharp increase in domestic prices, reducing levels of real consumption of the poor, who are net buyers of food. Similarly, a sudden change in import policy for cotton or sugar might put millions of farmers out of work if, in that year, world prices are low and there is a glut in production. It also had to be recognized that world markets in many agricultural commodities are highly imperfect and narrow. India's entry or withdrawal from these markets could have a substantial impact on world prices, as was the case in respect of wheat imports in the past. The fact that major producers, particularly the US and countries in the European Union, supported their exports through extensive subsidy arrangements further complicated the picture. For all these reasons, there was a case for moving cautiously towards the goal of liberalizing agricultural trade. Timing and selectivity were crucial. Agricultural policy changes should not be introduced in a bad year for domestic crops or at a time of rising inflation. Selectivity in regard to crops is also necessary. Trade liberalization first had to be introduced in regard to those non-food commodities which did not figure heavily in the consumption of the poor.

A viable framework for import policy had been established by then with the elimination of most quantitative controls and a progressive reduction in tariffs. One area where progress in liberalizing imports had been relatively slow was that of consumer goods.

## Import Policy for Consumer Goods

The question of liberalizing imports of consumer goods had figured prominently in recent public debate at that time. The government had taken some initial steps towards making imports of such goods easier by permitting some imports against special licences issued to exporters. However, on the whole, the policy remained highly restrictive. There were several issues involved including that of effects on domestic employment, which need to be considered before embarking on a bolder policy in this regard. In considering these issues, the experience of some other countries could have been useful:

- Between 1985 and 1992, Mexico virtually abolished all import controls, and lowered tariffs from an average of 23 per cent to 13 per cent. The ratio of imports to GDP increased from 8.2 per cent in 1985 to 14.6 per cent in 1992, and the share of consumer goods in total imports also increased from 7 per cent to 16 per cent. There was some evidence of reductions in firm-level employment, and a fall in real wages of workers.
- Between 1974 and 1979, Chile eliminated all import restrictions, and reduced tariffs from an average of 90 per cent to a flat 10 per cent on all imports. As a result, imports increased substantially as a proportion of GDP—from 10 per cent in 1971–73 to 22 per cent in 1982–85. The share of consumer goods in total imports also increased from 18 per cent to 27 per cent during the period. At the same time, exports also expanded rapidly and trade balance was positively affected by liberalization. As regards employment, manufacturing as a whole

experienced a decline, but this was offset by large gains in employment in agriculture and exports. Productivity growth in manufacturing improved and production in tradeable sectors became more labour intensive.

- Brazil had gone through two liberalization phases—from 1964 to 1974 and again from 1990 to 1994. The first phase came to a halt after the first oil shock because of BOP problems. Trade policies in the 1980s were uneven in view of the emergence of a debt problem. The overall framework remained restrictive with occasional episodes of liberalization. After 1990, Brazil again embarked on a major liberalization programme. However, this was reversed in 1995 with import restrictions being imposed on certain categories of durable goods (e.g., automobiles). The data on the impact of the recent phase of liberalization on growth of imports and employment in Brazil showed a mixed outcome and did not point to any clear conclusions. Considering that trade restrictions have had to be reimposed, it could be presumed that the recent experience had not been very positive.

- Finally, in the case of Korea, imports were fully liberalized over a 10-year period (between 1978 and 1988). The programme of liberalization was pre-announced and implemented gradually in order to avoid disruption and provide sufficient time to domestic industries to adjust to external competition. Imports of those goods which were believed to be fully competitive were liberalized first. Exports grew faster than imports, and no BOP problems occurred, except briefly after the second oil shock in 1979. There was a change in the composition of imports in favour of manufactured products (as against imports of oil and other primary commodities), but the share of consumer goods did not increase much. It increased from about 3 per cent of total imports in 1980 to 5 per

cent in 1990 despite removal of controls and lowering of
tariffs. Liberalization of imports was also accompanied by
a consistently high growth rate of output and employment.

The evidence from the experience of these countries is clear—a
generalization about the likely impact of import liberalization *per
se*, including liberalization of consumer goods, is not possible.
The impact can be good or bad depending on the circumstances
of a country and conditions under which imports are liberalised.
The experience in Chile and Korea was highly favourable.
The results of liberalization in Mexico were negative, and the
experience of Brazil was uneven. One lesson of the international
experience, which was relevant for India, was that the process of
liberalization of consumer goods has to be carefully 'managed',
as was the case in Korea. A phased elimination of controls and
reduction in tariffs over a pre-announced period has much to
commend itself. A gradual liberalization can have a positive effect
on employment and manufacturing production as it allows time
for firm-level adjustments, and for reallocation of resources in
line with a country's comparative advantage. A second important
conclusion was that the key to success lies in export growth.
Import liberalization of consumer goods should, as far as possible,
accompany or follow sustained increases in exports. This was
the policy adopted by Chile and Korea as well as several other
successful countries, including Japan.

In India's economic situation characterized by strong
export growth and high foreign exchange reserves, the case for
liberalizing imports of consumer goods over a period of three
to five years was strong. To begin with, quota restrictions should
be replaced by higher-than-average tariffs. Tariffs could then
be brought down according to a pre-announced schedule over
the next three or four years. That would allow uncompetitive
firms to adjust, while giving a strong push for fresh investment
in competitive firms. An advantage in India's favour was that

domestic prices of consumer goods, despite quota restrictions, were lower than world prices (except in sectors where past policies severely restricted competition and output growth, as in high technology electronics). The main problems were those of quality and durability. In a competitive environment, given sufficient time, these problems can be overcome with induction of better technology and improved management practices.

## Multinationals: Demons or Angels?

In India, till the mid-1940s, foreign capital dominated industrial and financial fields. The foreign trade network, as also part of the internal trade that fed into exports, was controlled by foreign capital. British companies dominated coal mining, jute industry, shipping, banking, insurance, and tea and coffee plantations. Moreover, through their managing agencies, British corporations controlled many of the India-owned companies. After 1920, the British giant companies—Unilever, Imperial Chemical Industries—were joined by several American multinationals, among them General Motors.

The large presence of foreign companies before Independence, however, did not contribute to the growth of income in the country. In fact, it may have been a cause of India's underdevelopment as foreign investment was concentrated in production and export of raw materials and foodstuffs. There was practically no transfer of capital to India and India was a net exporter of capital to the UK. There was no scope for transfer of technology as most of the investment was concentrated in low technology extractive industries.

Against this background, it is no wonder that after independence in 1947, an important plank of India's development policy was to discourage inflows of foreign capital. Foreign shareholding in existing companies was also reduced drastically by forced or voluntary transfer of capital into Indian hands. By the beginning of the 1980s, the share of FDI in gross capital formation was among the lowest for India amongst all developing countries (only 0.2 per cent as against the average of about 6 per cent for developing countries as a group). The highly restrictive policies towards foreign equity investment continued, without any significant change, until mid-1991. In the four years after that, rules governing foreign investment were liberalized greatly and India once again, after a gap of nearly 40 years, actively started seeking foreign investment.

Since 1970, there had been a dramatic transformation in the sectoral composition of both the flows and the stock of FDI. Foreign investment was no longer going into primary products or resource-based manufacturing. The concentration was mainly in services and technology-intensive manufacturing. The past investment of transnational corporations in extractive sectors (including petroleum and natural gas) had been substantially reduced through nationalizations. There were also strong interlinkages between flows of direct investment on the one hand and trade, financial flows and technology transfers on the other. In the case of the US, for example, at least 80 per cent of the trade was undertaken by transnational corporations. As regards technology transfers, in the 1980s, some 80 to 90 per cent of technology payments by developing countries to Germany and the US took the form of payments from affiliates to their parent companies and were directly associated with FDI.

The early 1990s witnessed equally dramatic developments in regard to net flows of private capital, including direct investment, to developing countries. Net flows were estimated to have increased to $175 billion in 1994, more than four times the 1989 figure of $42 billion. In volume, these were three times as high as official development assistance and other forms of official capital flows. During this period, FDI tripled in volume—from about $25 billion in 1980 to $78 billion in 1994.

Unlike in the past, a striking new development was that most of the jobs created by FDI were in developing countries. As many as 5 million out of 8 million new jobs associated with foreign investment were located in developing countries. The real wages of labour employed in these firms had also been rising faster than average incomes. Another feature of such investments was their strong presence in exports, with high domestic value added. A negative feature of foreign investment, from the point of view of recipient countries, was that it was more volatile than purely domestic investments. Capital tends to move out in response to increase in domestic wages over time or to political and other uncertainties.

In the 1990s, India was a relatively new destination for FDI, and the volume of such investment was still very low by international standards. Apart from its skills base, another advantage that India had over other developing countries was that of a relatively mature domestic industrial sector. As such, most of the new direct investment proposals were in the form of joint ventures with Indian partners, managed and staffed by Indian nationals or persons of Indian origin, which would ensure greater stability of investments in the future.

In view of the changes in India's economy in the preceding 50 years as well as changes in the nature of FDI, there was a strong case now for treating all foreign equity investment in the same way as domestic corporate investment, subject to two or possibly three exceptions. One clear exception was defence- and security-related industrial units (e.g., manufacture of armaments), where ownership and investment must be restricted to domestic corporations. Secondly, there may be certain industries or activities which employ a large number of persons, but which have become uncompetitive because of technological changes or shifts in comparative advantage. If entry of foreign companies in these traditional activities (or new substitutes) was likely to render a substantial number of persons unemployed, then there was a case for restricting foreign investment in these areas.

Some agro-based activities, which employ millions of persons throughout the country, would deserve protection from foreign investment on this ground. However, it was important to ensure that such cases were considered exceptional and did not become generalized to cover uncompetitive and technologically outdated units in the modern industrial sector, where the employment angle was not very significant, as in electronic hardware. A third exception, which was less clear-cut, was discrimination against foreign investment on 'cultural' grounds. A case could be made to prevent entry of, say, foreign films, or to restrict foreign investment in T.V. or radio stations in order to protect Indian culture and Indian values.

Although foreign investment policies had been liberalized in recent years, there was by no means unanimity of views among experts, and indeed the public, on the effects of such investment in India. Three different sets of arguments had been advanced against foreign investment. The first was the political argument that large scale of foreign ownership of industrial assets could pose a threat to national sovereignty or mortgage the national interest to foreigners (as indeed was the case during the colonial period). While it was conceded that this fear was far-fetched at that point in time, it was argued that it could become real in the future if there was a large flow of funds from abroad for a number of years. There was no rational way of discounting such fears about the distant future. The only practical way of preventing such a possibility from arising is to ensure that foreign investment was subject to domestic laws and domestic industrial policies. In view of the large size of the economy and the substantial volume of domestic investment, it was unlikely that foreign investment would ever become the predominant form of capital formation in India or exceed 'prudent limits'.

A second set of arguments against foreign investment concerned possible welfare-retarding cross-border transactions between the parent company abroad and the subsidiary company at home (e.g., 'transfer pricing' leading to transfer of capital abroad). This, in effect, was a widely used method of transferring capital during the colonial period. In the post-colonial period, in some countries, this method was also used to transfer profits by companies which were subject to price controls (e.g., drug companies) or companies which were not allowed to expand domestically (e.g., companies in the consumer goods sector). In recent years, the picture had changed. The world markets had become competitive and there was a struggle among all companies to maintain or increase their market shares. It was no longer in a company's interest to increase costs by resorting to transfer pricing, unless domestic markets were sheltered and protection levels were unduly high. In India, with the lowering of tariff levels and the removal of domestic investment restrictions, the temptation to raise costs artificially should be less. The correct answer to the problem of 'transfer-pricing', under the prevailing circumstances, was to encourage competition and further reduce barriers to new investment.

Finally, based on empirical research on the past performance of multinational companies in India, it had been suggested that their net contribution to the Indian economy had often been negative. Thus, it has been pointed out that these companies exported too little, that they tended to declare high dividends, that their investments were concentrated in certain sectors, that they transfered very little technology, and so on. Many of these findings about the past behaviour of these companies were no doubt correct. However, in this respect, their behaviour had been no different than that of domestic companies. If markets are highly protected and fragmented, if competition is actively discouraged, if there is a bias against exports and if there is no threat from new entrants, then industrial corporations, domestic or foreign, will behave in a predictable way in search of maximum profits with minimum work. The observed results in respect of past operational strategies of multinational corporations were, therefore, not surprising. Under those circumstances, the bias against foreign investment in India's policies was also understandable. However, the policy environment in the mid-1990s, both domestically and internationally, was very different. The past performance of foreign, or for that matter, domestic companies, was not an adequate guide to their future behaviour.

To sum up, FDI in India in the mid-1990s was too small to have economy-wide effects. As the volume of investment were to expand, it was necessary to ensure that other macroeconomic policies, particularly tariff and trade policies were such that positive effects of such investments were maximized. If protection levels were high and if there were trade restrictions, foreign investment could become a means of taking advantage of such high tariffs to generate high domestic profits. It could worsen the BOP if it resulted in large imports of components, raw materials and capital equipment from the parent company without adequate exports. Similarly, special fiscal incentives, which were not available for domestic investment, also had to be avoided. Such special incentives may simply encourage 'round trip' capital flows from the host country, without adding to total savings or investment. The best way of encouraging FDI was to implement more open trade policies and to improve the environment for all investments, including domestic investment.

## Management of Capital Flows

International research on management of a policy for capital flows has broadly centred around three questions: (a) how to 'prevent' the occurrence of excessive capital flows; (b) how to avoid the adverse effects of excessive capital flows on domestic economy in case they do occur; and (c) how to manage a foreign exchange crisis from getting out of hand, in the event there is a collapse, for whatever reason, of confidence in the economy.

Between the mid-1980s and mid-1990s, certain important

changes took place in the scale and composition of external capital flows. From a level of $42 billion in 1989, net capital flows to developing countries increased to $175 billion in 1994. A dramatic change also occurred in the composition of these flows. Earlier, most of the capital flows were in the form of direct foreign investment or loans from international banks. In later years, however, the main sources of external financing in many developing countries were international bond and equity issues. Both these sources of funds are generally more 'liquid' than direct investment in plants and syndicated loans from banks. Another characteristic of bond or equity issues is that they are generally more widely held than bank loans. Bonds are spread among a large number of investors, including mutual funds. They are also more actively traded.

In view of these characteristics (i.e., liquidity, widespread holding and active trading), bond and equity markets tend to be more volatile and subject to 'speculative' movements than, say, syndicated loans from banks. Trading in bonds and equities in large foreign markets, such as those of the UK or the US, is also less subject to prudential supervision than bank lending. Rumours about future policy outlook or misleading information about behaviour of other players in the market can easily start a panic. Panic-driven activity in the bond or equity market is also less easy to control in the absence of official intervention and other safety nets. The sheer volume of international financial transactions and instant communications among distant markets further complicate the task of managing a crisis once it occurs. In 1993, the notional value of global 'over-the-counters' and exchange traded derivative financial products alone exceeded $12 trillion, about twice the size of the nominal GDP of the US.

A more workable idea for developing countries was to not rush into full convertibility of capital account transactions until the domestic financial system was sufficiently strong to cope with volatility in capital flows and there was reasonable macroeconomic stability. It is generally accepted that excessive capital flows can create difficult problems for the management of the domestic economy. Both a tightening of monetary policy and an appreciation of the exchange rate may become unavoidable, which in turn may affect output and exports. Excessive flows may also bring about a banking crisis, as it exposes banks to rapid expansion of their loan portfolio as well as to market risks—exchange rate and interest rate risk—that cannot be fully hedged. In case fiscal and monetary policies are tightened, banks may face further problems because of loan defaults.

The most widely used policy intervention to restrain the growth of money supply and appreciation of the nominal exchange rate is to 'sterilize' the domestic currency counterpart of excess capital inflows. This is done either by raising the cash reserve requirements (against deposits) or through open market operations in government securities. While sterilization is usually effective in the short run to take care of a temporary upsurge in capital flows, it can have several negative side effects in the medium or long term. Increase in reserve requirements (or, for that matter, open market operations) can lead to an increase in domestic interest rates. This widens the domestic-international interest rate spread, and may lead to higher inflows of short-term capital. Higher interest rates on government securities may also add to the pressure on the budget. It is, therefore, preferable to adopt 'preventive' policies to discourage destabilizing excess flows, rather than to bury them after they have arrived.

It seems that, for a developing country with limited access to exceptional financing, a good option is to regulate certain types of volatile capital inflows and prevent crisis from arising rather than count on international support in an emergency. It is, of course, not feasible to derive universally valid policy guidelines for capital flows, either in theory or in practice, which can fit all countries at all times. In the context of India in the mid-1990s, and based on the international experience so far, it was possible to suggest the following broad framework of policy which would avoid some of the risks inherent in such flows, and maximize the potential gains:

- A fundamental point is that a large trade or CAD cannot persist for more than a few years, and in any case cannot be financed by private capital flows from abroad. Despite the large volume of such flows globally, most of the domestic savings of capital surplus countries are in fact invested domestically. It is difficult to prescribe an upper limit for the level of sustainable current account deficits; however, there is hardly any country which has been able to sustain a deficit of more than 3–4 per cent over a number of years (the exceptions are countries which are supported by large flows of bilateral aid for political reasons).

- The best form of private capital flows is direct investment in plants and factories. Such investments are less likely to be withdrawn, even though the annual volume of fresh inflows may fluctuate. However, it is important to ensure that such flows are not of a 'tariff-jumping' variety. A reduction in domestic protection and tariff levels, therefore, has to be an integral part of the policies for attracting direct investment.

- Long-term commercial debt, while useful for investment financing, should be kept within prudential limit of debt-servicing capacity. For India, a safe rule is that, for any length of time, debt servicing on all external debt should not exceed 20 to 25 per cent of receipts from exports and services.

- A developing country wishing to attract portfolio capital is well advised to maintain a high level of external reserves. A good rule is that foreign exchange reserves should be at least twice as high as the total stock of portfolio and

short-term investments. In a panic, this should provide
sufficient resources to cover withdrawals of capital as
well as payments for debt service and imports. In fact,
if reserves increase *pari passu* with increase in portfolio
investments, a panic is unlikely to occur. 'Panics' are
self-fulfilling in the sense that once markets expect a
payment crisis, the crisis will occur, as all investors rush
to withdraw funds. Industrial countries, which have
open financial markets, can manage with lower levels
of reserves because of 'safety net' arrangements among
their central banks, and their mutual facilities to borrow
in foreign markets. Developing countries do not have
adequate access to such arrangements.

- As a temporary device, sterilization of capital inflows
  may be unavoidable from time to time. If capital flows
  are persistently larger than those required to finance
  sustainable CADs and additional foreign reserve
  requirements, then capital controls on short-term and
  portfolio capital inflows should be introduced. Such
  controls should be transparent and simple to administer.
- The health of the domestic banking system should not
  be allowed to become tied to the ebb and flow of foreign
  capital. Regulations that limit the exposure of banks to the
  volatility in equity and the real estate markets would help
  insulate the banking system from the bubbles associated
  with large capital flows.
- Forex markets dislike 'surprises' even more than trading
  losses. The impact of the latter is known, and can be
  provided for, while a surprise change in policy can easily
  create panic (in anticipation of more surprises). Therefore,
  the policy for capital transactions should be clear and
  stable. In any case, controls on outward flows of capital
  held by foreigners should be scrupulously avoided, unless
  they are part of a pre-announced policy. The government

should, however, always reserve the right to regulate inward flows if these become excessive. This should be announced as part of the policy package governing capital flows.

# 4

## INTERNATIONAL FINANCIAL ARCHITECTURE

### 2002

In the 1990s, the subject of a new international financial architecture had been a matter of considerable debate among governments and central banks all over the world. This essay deals with certain issues concerning the international financial system from the perspective of developing countries, particularly India.

### Why the Debate?

The financial crisis that occurred in East Asia after July 1997 generated enormous international concern over the global financial stability. This concern translated itself into a debate on how to prevent such crises from arising in future and how to manage them, if they do arise. Naturally, such a debate included a review of the appropriateness of existing international financial arrangements and the need for redesigning such arrangements. Various facets of relevant reforms considered after the crisis came to be referred to as the International Financial Architecture (or IFA, for short). At the outset, it may be useful to recall the reasons for considering a new IFA in the context of the crisis.

First, the crisis of 1997 was sudden and unanticipated. Neither the countries themselves nor the credit rating agencies or the international financial institutions anticipated a crisis of such magnitude. Second, the speed with which the crisis got transmitted to other countries took the financial community by surprise. New technology made capital flows easy and fast; it also made the transmission of panic easy. Third, there was evidence of failure of market mechanisms as was generally understood, necessitating a review of relative roles of the State and the market in financial systems. Fourth, when the failure of markets resulted in systemic threats, the burden of market failure got shifted to governments and the public sector. This burden was borne, to a large extent, by the governments of borrowing developing countries, rather than lending or by developed financial centres, though 'irrational exuberance' in lending and borrowing are inseparable. Fifth, although the international financial institutions, in particular the IMF, came to the rescue of the affected countries, there were questions regarding the timing, content and adequacy of their assistance package. Sixth, the contagion spread across and beyond the region to developed countries also. Global interdependence was no longer merely a slogan or one-sided affair. It became clear that developments in developing countries were important for stability in the entire financial world.

The experience with the Asian crisis led to the perspective that international financial architecture as it existed, was inadequate to meet the new realities. It started with the expression of dissatisfaction by several experts and some countries on the role of credit rating agencies, and to some extent, on the role played by the IMF. As the debate proceeded, there occurred a distinct shift in emphasis in favour of considering microeconomic aspects in addition to macro-policies, such as fiscal and monetary management. There was also a shift from abstractions to institutional aspects such as corporate governance, regulatory structures and transparency. Some attention was also given to mechanisms whereby the government could ensure market accountability, and not merely market efficiency, lest the burden of market failure shifted entirely to governments. Finally, the debate had to recognize that the divide between developed and developing countries or the OECD and non-OECD countries was not as meaningful as it was thought to be. The developing countries themselves represented a wide range with different degrees of openness of their economies, and consequently different degrees of integration with the global financial system.

## Levels of the Debate

The debate on IFA occurred at different levels and in different fora. It occurred within the international financial institutions, particularly the IMF and the World Bank. Discussions were also held in the United Nations. Intellectuals from both developed and developing countries expressed their views on various aspects of the IFA and exercised influence on the direction and the course of the debate. The private sector in major financial centres participated in the debate, particularly through the Institute of International Finance. The voluntary international organizations consisting of national-level regulatory and supervisory bodies, such as the International Organization of Securities Commission and the International Association of Insurance Supervisors, took this issue on board. Policy discussions were also conducted under the auspices of the traditional groupings, such as the Group of Seven (G-7) of developed countries and G-24 of developing countries.

The debate within these different fora and at different levels contributed to structured and formal discussions in the Development Committee and the Interim Committee, subsequently renamed the International Monetary and Financial Committee, respectively, especially the latter, which is formally charged with the responsibility of guiding the international community in matters relating to global economic cooperation. The discussions in these formal groupings were aided by informal and expanded high-level official meetings of developed and developing countries. Among these, a mention may be made of the US-sponsored G-22, and G-33, which is sponsored by G-7. These resulted in another group, known as the Financial Stability Forum, being set up in February 1999.

Several seminars of G-33 on the IFA were convened at the initiative of the finance ministers and central bank governors of G-7. G-20 was formally established in September 1999 which superseded G-33, and comprises G-7, 11 other countries (including India) and two institutional representatives. G-20 promotes consistency and coherence in the various efforts aimed at strengthening the international financial system, and addressing issues that go beyond the responsibilities of any one organization. India has been involved in almost all the discussions where developing countries have been represented, and an effort has been made by India's representatives as well as representatives of certain other developing countries (including Brazil, China, Mexico, Russia, Saudi Arabia and South Africa, who are also represented in G-20) to present the developing country's perspective on various issues.

## Main Issues in the Debate

The debate on IFA covered a wide spectrum of policy issues. Issues of particular significance to developing countries were the exchange rate, policy on reserves, the role of the external private sector, management of capital flows, strengthening the financial system, transparency codes and standards, reform of international institutions and new arrangements for international liquidity.

## The Exchange Rate

Appropriateness of the exchange rate regime is considered to be one of the major factors relevant to the Asian crisis. Even at the time of the crisis, this debate was not new and there were, as always, differing views on an appropriate exchange rate regime. One view was that a fixed exchange rate can promote domestic macroeconomic and financial stability by providing a firm nominal anchor. However, it was pointed out—and more sharply so after the Asian crisis—that there are risks associated with pegged exchange rates. This was reflected in the fact that the East Asian countries with pegged exchange rates were the worst affected.

A widely expressed view at the time was that countries should eschew variable peg regimes in favour of either something much higher or the voluntary adoption of 'managed floating'. Currency board was an option that has been favoured by some developing countries, particularly small open economies. In reality, however, the actual policy adopted by most central banks was different from the theoretical optimum. The most common exchange rate regime adopted by countries, including industrial countries, was neither a currency board nor a free float. Most of them had adopted intermediate regimes of various types, including fixed pegs, crawling pegs, fixed rates within bands, managed floats with no pre-announced path, and independent floats with foreign exchange intervention moderating the rate of change and preventing undue fluctuations. Largely, barring a few, countries had 'managed' floats or central banks intervened periodically.

There had been some debate on the policy measures that could be taken by the central bank to keep exchange rate movements 'realistic' and relatively orderly. An analysis of the behaviour of the market during the 1990s and beyond showed that the day-to-day movements in the exchange rate in the short run had little to do with the so-called fundamentals or the country's capacity to meet its payment obligations, including debt service. Any adverse news and expectations generated by it tended to play a paramount role in generating self-fulfilling expectations. Developing countries in general had smaller and localized forex markets, where normal domestic currency values were expected to show a depreciating trend, particularly if relative inflation rates were higher than most major industrial countries. In this situation, there was a greater tendency among market participants to hold long positions in foreign currencies and hold back sales when expectations were adverse and currencies were depreciating rather than the other way round, i.e., when currencies are appreciating and expectations are favourable. Thus, market behaviour was not symmetrical in both directions.

There was a fair degree of agreement that stability in the exchange rate is well served by the stability in the conduct of monetary policy. An increasing number of central banks, particularly in industrial countries, were directing monetary policy to the sole objective of price stability. Countries were also increasingly announcing near-term inflation targets explicitly. Transparency in objectives, intermediate targets and operating instruments was expected to play an important role in anchoring expectations.

However, in spite of cautious policies, if there were unanticipated or unacceptable movements in exchange rates, it was also generally, but by no means unanimously, held that central banks must be prepared to move interest rates in order to stabilize expectations. In the situations, direct intervention in forex markets by the central bank could also become necessary, although the precise extent and modality of intervention was likely to depend on the individual country's circumstances, especially the size and depth of the market and the size of the country's foreign exchange reserves. It had been recognized that developed countries, and in particular the central banks of major financial centres, were in a position to intervene in the exchange market on behalf of each other. However, such options were not available to most developing countries. This imposed additional challenges to the central banks, both in tactics of intervention in the forex markets and the need for supplementary monetary and administrative measures.

India's exchange rate policy had evolved from the rupee being pegged to a market-related system (since March 1993). By the 2000s, the exchange rate was largely determined by the market, i.e., demand and supply conditions. The objective of exchange rate management had been to ensure that the external value of the rupee was realistic and credible as evidenced by a sustainable CAD and manageable foreign exchange situation. Subject to this predominant objective, the exchange rate policy was guided by the need to reduce excess volatility, prevent the emergence of destabilizing speculative activities, help maintain adequate level of reserves and develop an orderly foreign exchange market.

The Indian market, like other developing countries' markets, was not yet very deep and broad, and was characterized by an uneven flow of demand and supply over different periods. The market was also characterized by a few major players, and lumpy public-sector demands, particularly on account of payments for oil imports and servicing of public debt. In this situation, the RBI had been prepared to make sales and purchases of foreign currency in order to even out the lumpy demand and supply in the relatively thin forex market and to smoothen jerky movements. However, such intervention was not governed by a pre-determined target or band around the exchange rate. While it was not possible for any country to remain completely unaffected by developments in the international exchange markets, fortunately India had been able to keep the spillover effect of the Asian crisis to a minimum through constant monitoring and timely action, including recourse to strong monetary measures, when necessary, to prevent the emergence of self-fulfilling speculative activities.

India's exchange rate policy of focusing on managing volatility with no fixed rate target while allowing the underlying demand and supply conditions to determine the exchange rate movements over a period in an orderly way, had stood the test of time. Despite several unexpected external and domestic developments, India's external situation continued to follow the same approach of watchfulness, caution and flexibility in regard to forex market.

## International Reserves

Another issue that had figured prominently in the existing debate on forex management was the question of appropriate policy for management of foreign exchange reserves. In a regime of free float, it could be argued that there was really no need for reserves. Some countries, where monetary policy is directed towards the single objective of inflation control, in fact, maintained very small reserves, essentially for operational purposes. In the light of the volatility induced by capital flows and the self-fulfilling expectations that this could generate, there was then a growing consensus for emerging market economies to maintain 'adequate' reserves. The emerging economies had to rely largely on their own resources during external exigencies (or contagion), as there is no international lender of last resort to provide additional liquidity at short notice on acceptable terms. Thus, the need for adequate reserves was unlikely to be eliminated or reduced, even if exchange rates were allowed to float freely.

Traditionally, the adequacy of a country's reserves had been mainly linked to import requirements. Recent experience had, however, shown that countries which were holding large levels of foreign currency reserves in relation to imports, did not necessarily escape the crisis. These countries had large reserves, but these reserves disappeared quickly as they tried to defend their currencies. Some part of the reserves could not be accessed when it was most required by these countries since they were invested in illiquid assets such as real estate, stocks with low trading volume or collectibles. Further, there was no transparency with regard to the precise figure of unencumbered reserves.

In the aftermath of the Asian crisis, the emphasis had shifted from measuring the adequacy of foreign exchange reserves only in relation to imports to measuring the usable or unencumbered reserves in relation to short-term liabilities, in particular short-term debt. On this subject, a notable suggestion for emerging market economies was that countries should manage their external assets and liabilities (and the so-called 'liquidity at risk') in such a way that they are able to live without new foreign borrowing for up to a year. In other words, usable foreign exchange reserves should exceed scheduled amortization of foreign currency debts (assuming no rollovers), in addition to the estimated CAD, during the following year. This has implications both for the level of reserves and debt management since this implies a limit on the size of the debt, in particular short-term debt and debt that is falling due for repayment.

The link between short-term debt and reserves came to the fore in the IFA deliberations. There is a cost to building up reserves through large debt flows since the cost of debt would generally be higher than return on reserves. At the same time, a high level of reserves satisfies the need for liquidity, offers insulation against unforeseen shocks and acts as a source of comfort to foreign investors. The essence of reserves management being safety and liquidity, it stands to reason that all investments made out of reserves should be of top credit quality and excellent liquidity. Hence, the return on reserves and the cost of borrowing are not strictly comparable.

India has been steadily building up reserves by encouraging non-debt creating flows and de-emphasizing debt creating flows, particularly short-term debt. In fact, this strategy, coupled with the maintenance of an acceptable level of CAD and market-determined exchange rate regime were cornerstones of the policy of external sector management recommended by the Report of the High Level Committee on Balance of Payments (Rangarajan Committee) in April 1993.

In the context of the changing interface with the external sector, and the importance of the capital account, reserve adequacy was now evaluated by the RBI in terms of several indicators and not merely through conventional norms, such as the import cover. The overall approach to the management of India's foreign exchange reserves is reflected in the changing composition of the BOP, and endeavoured to reflect 'liquidity risks' associated with different types of flows and other requirements. The policy for reserve management was thus judiciously built on a host of identifiable factors and contingencies. Such factors *inter alia* included the size of short-term liabilities, the possible variability of portfolio investments and other types of capital flows, the unanticipated pressures on BOP on account of external shocks or increase in oil prices and movements in repatriable foreign currency deposits of NRIs.

As on 18 January 2002, India's foreign exchange reserves were about $49.21 billion, which included $2.85 billion of gold and $46.36 billion of foreign currency assets. India's foreign exchange reserves were only $5.8 billion in March 1991. The previous five years had seen increases in reserves of the order of $4.7 billion in 1996–97, $2.9 billion in 1997–98, $3.1 billion in 1998–99, $5.5 billion in 1999–2000 and $4.2 billion in 2000–01, i.e., a cumulative increase of $20.4 billion despite the East Asian crisis and a number of other domestic and external pressures on the foreign exchange situation.

India's policy of building and maintaining adequate level of reserves while at the same time constraining debt, especially short-term debt, has served well during the Asian crisis and subsequently as well. In the light of historical experience during the 1960s and '70s, it was also apparent that a relatively high level of reserves had also contributed to enhancing India's national and economic security during periods of global and regional uncertainties (for example, the post-Kargil period, and the period following the imposition of US sanctions).

## Involvement of the External Private Sector

By the early 2000s, the role of external private-sector lenders and banks in forestalling and resolving a debt crisis, and more broadly in external liability management, had started assuming importance in the context of a much larger movement of private capital as compared to the flow of resources from the official sector. Efforts to involve private-sector lenders in liability management has many objectives, viz., to bring about a more orderly adjustment process, limit moral hazard, strengthen market discipline, and help protect countries against volatility and contagion. The key issue concerning the role of private-sector lenders in forestalling and resolving the crisis was whether by involving the private sector, the overall costs associated with foreign exchange crisis could be reduced, either by smoothening the crisis-resolution process or by reshaping the incentives under which private institutions operate. The idea was to proactively institute mechanisms in place before any crisis took place, so that the resolution of the crisis was more orderly.

Resolution of a crisis by involving the private sector requires addressing the balance of burden-sharing between official and private creditors and between debtor and creditor countries. Some experience had been gained from recent efforts to secure private-sector involvement in the resolution of crises. Agreements for the maintenance of exposure on short-term bank credit had been achieved both voluntarily and through the application of moral persuasion by central monetary authorities. In addition, international sovereign bonds had been restructured through voluntary debt exchanges.

A broad consensus had also emerged among member countries of the IMF on the need to seek private-sector involvement in the resolution of crises, while providing for flexibility in the form of involvement and in the methods used to ensure it. The IMF had made some progress in developing a framework for crisis resolution based primarily on the Fund's traditional catalytic role in lending by persuading private creditors to maintain exposure. In extreme cases, i.e., when the financing requirement is large and the member has poor prospects of regaining market access in the near future, or if the member has a debt burden that appears unsustainable in the medium term, the Fund can also urge private creditors to agree to a voluntary standstill clause in loan contracts with the debtor country.

The success of the Fund's involvement in crisis resolution by involving the private sector had been limited till then. Since the Fund had not established explicit guidelines on the stipulation of a voluntary standstill clause, the Fund and its major shareholders enjoyed considerable discretion about the intervention process. There was, in any case, a strong opposition from some major players, and from participants in private markets, to mandatory mechanisms for 'binding in' and 'bailing in' the private sector. Further, in some of the recent examples of renegotiated debt settlements, creditors had not borne the consequences of the risks they had taken. On the contrary, they had forced the governments of developing countries to assume responsibility of private debt and accept a simple maturity extension at penalty rates. In the absence of a statutory protection of debtors, negotiations with creditors for restructuring loans could not be expected to result in equitable burden-sharing.

While there was a consensus regarding the need for the private sector to share the burden, discussions in various fora had centred on the 'mode' of burden-sharing. Ideas being considered included contingent credit lines, embedded call options, debt-service insurance, bond covenants, bankruptcy procedures, debt standstill and creditor-debtor councils. A device providing a mechanism of burden-sharing was by taking recourse to the principles of orderly debt workout along the lines of the United States Bankruptcy Code. This proposal was primarily aimed at addressing financial restructuring rather than liquidation. It allowed temporary standstill on debt servicing, access to working capital and reorganization of assets and liabilities of the debtor.

Basically, these mechanisms would bind private-sector participants to either provide additional funds, or reduce debt service burdens in times of crisis without creating moral hazard or disrupting normal market conditions. For instance, it had been suggested that countries could contract market-based contingent credit lines with commercial banks to trigger liquidity support in times of crisis. If the credit lines were fairly priced, they could provide effective insurance against adverse market developments. Similarly, it had been proposed that standstill arrangements, which allowed the private sector to extend additional credit when there had been default on existing debt, should also be considered. Such mechanisms were considered essential since the absence of such arrangements made it possible for banks holding short-term inter-bank claims to leave the market unscathed or significantly reduce their exposure, which contributed to a perverse incentive structure, inequitable burden-sharing and panic. Creditor-debtor councils could serve to improve the flow of information. Similarly, bond and debt covenants could introduce sharing clauses and provisions for the modification of terms by qualified majorities in order to speed up the negotiation process.

From the official sector, the IMF has decided to establish a contingent credit line (CCL) as a precautionary mechanism to ward off financial crisis. This was expected to supplement and not substitute the contingent credit lines from the private sector. This support was to be available to members who were undertaking sound macro-policies but are affected by 'contagion'. The CCL was expected to signal to the market the confidence of the IMF in the country's policies and boost confidence in the market.

It remained to be seen how flexibly the eligibility criteria in the new IMF facility would be operated in practice. For the success of the scheme, it seemed necessary to ensure that performance criteria were in fact appropriate and objective, so that there was no room for subjectivity or the introduction of political and other non-economic considerations. The scheme was likely to succeed in avoiding potential problems only if it is operated in a more or less automatic way subject to certain pre-determined quantitative and objective criteria being satisfied by the country. For example, one such test could be the extent of outstanding short-term liabilities in relation to a country's net reserves. Another test could be a pre-specified and an agreed level of fiscal deficit or the level of aggregate demand. As long as a country was following policies that satisfied these and similar quantifiable criteria, it was expected to be able to use this new facility without further ado.

## Management of Capital Flows

After the Asian crisis, there had been an extensive debate on the issue of capital account liberalization and controls. The issues being debated related to the desirability, form and content of capital controls, risk containment strategies in external debt management and the desirable sequencing of capital account liberalization.

It was widely agreed that a major source of vulnerability in the Asian crisis was the accumulation of short-term liabilities of banks and corporates and poor quality of risk assessment. Overall, there appeared to be a consensus in favour of developing countries restraining the inflow of short-term capital. There was, however, a growing debate on the means by which short-term flows could be controlled. Important suggestions related to effective monitoring, imposition of Tobin Tax, which is a tax on spot transactions, and unremunerated reserve requirements.

There had also been a suggestion to impose an increased capital requirement on inter-bank transactions to bring about greater discipline on cross-border, inter-bank market. It is self-evident that just as there is no irresponsible borrowing without irresponsible lending, there is no short-term borrowing without short-term lending. Unfortunately, in practice, the existing policy was biased in favour of short-term bank loans rather than medium- or long-term loans to developing countries. Thus, by and large, short-term loans to a country were likely to enjoy a higher credit rating than longer-term loans, irrespective of the overall record of a developing country in meeting its debt service obligations.

Basically, two positions significant to developing countries seemed to have emerged from the discussions on cross-border capital movements. First, while reiterating the longer-term efficiency of relatively free capital movements, a case had been made for capital controls either as a precautionary measure or as a temporary measure in a crisis. This was based on the view that financial markets were characterized by volatility and irrational expectations. Second, it was felt that emerging markets should not liberalize capital account in a hurry, without prior action for strengthening their financial systems. Financial markets are prone to rapid changes in perceptions and expectations because of poor information and herding. In developing countries, these are also typically thin and volatile. There was also a developing consensus that controls, when necessary, should be imposed on inflows rather than outflows and, as far as possible, they should be focused on short-term and volatile flows.

A related issue was the possible amendment of the IMF's Articles to extend its jurisdiction over the Capital Account. The IMF itself had acknowledged that there was no unique path that defined orderly capital account liberalization. In fact, the path and speed at which countries could traverse across this path would depend to some extent on the safeguards that the new international financial architecture could provide. It was worthwhile to note that a significant amount of liberalization of capital account had already taken place in the developing countries under the existing Articles.

From the viewpoint of developing countries, it needed to be recognized that external capital has benefits and acts as a complement to domestic savings. However, short-term reversible flows can also have negative effects on the economy, particularly during periods of political or economic uncertainty. While large inflows pose policy dilemmas for macro-management, large and sudden outflows can impose extensive damage to the financial sector and also result in a disproportionate output loss. It must be understood that merely by lifting all capital controls, the markets of a developing country do not get as deeply integrated as a developed country's markets. As such, each country would need to decide on its own path of capital account liberalization with regard to the timing and sequencing. This, in turn, is likely to depend upon the extent of stability and institutional structure of the domestic financial sector. Taking all these factors into account, it did not seem necessary to embark on a time-consuming procedure for the amendment of IMF's Articles in order to promote capital account liberalization. At that point in time, a constructive consultative process between the IMF and member countries on these issues was likely to be adequate to achieve the objectives in view.

India's policy on the capital account had been predominantly influenced by the recommendations of the Rangarajan Committee mentioned earlier. The committee's recommendations with regard to discouraging short-term flows stood vindicated by the prevailing international consensus on the need for controlling short-term debt. In India, short-term debt is carefully monitored with differential treatment between trade- and non-trade-related debt and is subject to a quantitative ceiling. Deposits by NRIs are also controlled through specification of interest rates or interest rate ceilings for different maturities. As a result of these efforts, the average maturity of external commercial borrowings in India was about six to seven years.

India had managed its capital account to ensure growth with

stability, while consistently adding to its foreign currency reserves. There were periods of capital surge during 1993–95, and also three major episodes of volatility in flows in the second half of 1995–96, during 1997–98 and again in August 2000. During periods of exchange rate volatility in the wake of the East Asian crisis, there were major imponderables involved, both externally and internally, and contagion and herd behaviour had to be guarded against. In these situations, a coordinated policy framework and careful calibration of policy instruments resulted in an effective management of capital flows without intolerable shocks on the performance of the economy.

## Strengthening the Financial System

In the debate on the IFA, increased recognition had been given to three prerequisites for the efficient functioning of the financial sector: a well-designed infrastructure, an effective market discipline, and a strong regulatory and supervisory framework. A well-designed infrastructure has many elements. First, it requires a proper legal and judicial framework. A second element relates to fostering good corporate governance. A third element is comprehensive accounting standards and a system of independent audits. The International Accounting Standards Committee had done substantial work related to this aspect. A fourth aspect relates to an efficient payments and settlement systems. Core principles had been established in this area also.

Effective market discipline also requires a good credit culture and well-developed and functioning equity and debt markets with a wide variety of instruments for risk diversification. The Basel Committee had revised its existing capital adequacy framework guidelines. The Committee has proposed a new approach for the risk weighting of book assets which offered a choice between a standardized approach based on ratings of external agencies and an internal ratings-based approach, in which risk weights are assigned using statistical techniques based on the probability of default. The Asian crisis had demonstrated that problems can arise when excessive leverage is coupled with excessive concentration of risk. Improving the assessment and management of risks was, therefore, another area that required attention.

International consensus had already been reached on what constitutes sound practice in many areas of banking supervision and securities regulation. The Basel Committee had released the Core Principles for Effective Banking Supervision and the International Organization of Securities Commissions (IOSCO) had produced similar guidelines for securities industry. Individual countries themselves were taking steps to develop markets, review their ongoing regulatory and supervisory procedures, and adopt international best practices. Countries were making efforts to strengthen the legal environment in which the financial systems were operating. Appreciable steps had also been taken to improve bankruptcy procedures.

Supervision of non-bank financial companies (NBFCs) was also conducted by the RBI. The prudential norms applied for the NBFCs closely resembled those for banks' capital adequacy. Asset classification, provisioning and income recognition norms had also been prescribed. Both, off-site and on-site supervision methods had been adopted and CAMELS (capital adequacy, assets, management capability, earnings, liquidity, sensitivity) pattern was used to rate and evaluate NBFCs. All NBFCs were required to publish their audited balance sheets.

There was a lot more work that needed to be done to make the financial system deeper and more vibrant. A number of committees had been set up to advise the RBI and the government on further steps. A Technical Advisory Committee on Financial Markets had been established as a consultative forum to advise the RBI on an on-going basis. A government committee has also looked into the legal and judicial aspects.

## Transparency in Systems and Developing
## Standard Codes

During the ongoing debate on IFA, an influential view was that the information made available to the markets by the official sector or by corporates or even by financial intermediaries, prior to the crisis, did not reflect realities in the emerging economies. The practices of disseminating information as well as its reliability, timeliness and quality varied sharply from country to country. Hence, considerable attention was devoted to developing uniform transparency codes and standards. For instance, IOSCO has coordinated with other agencies, such as the International Accounting Standards Committee, to develop proper accounting and disclosure rules for the securities market. A noteworthy feature of the exercise was that these efforts encompass the private sector as well, as many of the standards, e.g., accounting, auditing, bankruptcy, corporate governance and securities market regulation required to be implemented at the corporate level.

The IMF also developed voluntary standards in certain areas of the financial system. The Code of Good Practices on Fiscal Transparency was approved in 1998 and later revised in 2007. The Code of Good Practices on Transparency in Monetary and Financial Policies was also developed which was adopted by the interim committee (in 1999). The transparency policies listed in this Code were purported to focus on clarity of roles, responsibility and objectives of central banks' processes for formulating and reporting of monetary policy decisions, public availability of information and accountability.

It is, however, important to recognize that while transparency is of paramount importance, as it enables and improves the understanding of the stance of policies by market participants, the quality and content of transparency has to be appropriate and in tune with country circumstances. Given the divergence in institutional development and the nature of relations between various arms of national governments, it seemed unlikely that a uniform code could be universally valid at that stage. It stands to reason that the emphasis should be on voluntary adoption and gradualism rather than a 'big bang'. It was also important that the manner in which these international standards were monitored did not degenerate into categorizing countries as performers and non-performers. The goals of transparency could be best served by a balanced and symmetric evolution of information as between the authorities and the market participants.

So far as India was concerned, consistent with democratic traditions and free press, an impressive degree of transparency in provision of information was already in place. The RBI provided up-to-date weekly data on all relevant macroeconomic and financial indicators through the 'Weekly Statistical Supplement' to the RBI Bulletin. In addition, the RBI and the government disseminated as much information as possible through monthly, quarterly and annual publications.

Commercial banks in India were also required to maintain disclosure standards at par with those of international banks. This has been achieved by mandating disclosure of some of the essential strength indicators and performance-related parameters as part of commercial banks' balance sheets. In fact, since April 1997, even before the onset of the Asian crisis, Indian banks had been disclosing capital adequacy ratios (both Tier I and Tier II) separately, percentage of non-performing assets (NPAs) to net advances, provision made towards NPAs, and gross and net value of investments. Since March 1998, banks in India had been asked to disclose further information regarding interest and non-interest income as a percentage of working funds, operating profit as a percentage of working funds, and information on the financial position of subsidiaries. With effect from April 2000, the RBI had also advised banks to disclose maturity profile of loans and advances, investments and lending to sensitive sectors.

## International Financial Institutions

The discussions on the future shape of international financial institutions revolved round three pillars: strengthening the existing financial institutions, creating new institutions and establishing new groupings. There was consensus that international financial institutions (IFIs) must adapt to the changing environment if they were to maintain their effectiveness. Some general agreement on the principles that should guide efforts to enhance the effectiveness of IFIs had also been reached. These included arrangements to enhance accountability, an inclusive process for a broad range of countries and other institutions, and a more participative and open decision-making process.

The debate on strengthening the role of international financial institutions had also covered the role of rating agencies. The ratings given by these agencies seemed to have greatly influenced the decisions of investors in Asia and after the crisis, questions had arisen on the public accountability of rating agencies. An issue being debated was whether ratings should be autonomous or should rating agencies themselves be regulated. This assumed critical importance in the context of the consultative paper of the Bank for International Settlement (BIS) on the New Capital Adequacy Framework, which envisages the possibility of assigning risk weights according to ratings as an option available to banks.

A proposal was put forward to create a new international institution called World Financial Authority (WFA) or a Board of Overseers of major international institutions and markets with powers for oversight and regulation globally. Various models were envisaged for the proposed WFA. One proposal entailed the establishment of a body, which would have the responsibility of setting regulatory standards for all financial enterprises—funds and insurance companies as well as banks, and off-shore and on-shore entities. National regulators would remain responsible for implementing standards promulgated by the WFA. Another model was for the WFA to serve as an umbrella organization into which the existing bodies could be brought together. It was, however, well recognized that setting up such an institution would be a complex process. As an intermediate step, a suggestion was made in the Development Committee for setting up a new and Permanent Standing Committee for Global Financial Regulation. This intermediate proposal sought to bring together not only the World Bank and the IMF, but also the Basel Committee and other regulatory groupings on a regular basis.

These proposals, however, did not find much support. As a response to these proposals, the major industrialized countries set up the Financial Stability Forum with representatives from the finance ministries, central banks and regulatory authorities of G-7 countries, as well as from the IMF, World Bank, Basel Committee, IOSCO, International Association of Insurance Supervisors, BIS, OECD, Committee on Global Financial System and Committee on Payment and Settlement System. It was decided that the forum should be broadened by including participants from other industrial countries and emerging economies in order to make it more effective. The G-20 was formed in supersession of G-33 and included representatives from developing countries. Within the IMF itself, the interim committee was rechristened the International Monetary and Financial Committee.

## New Arrangements

Finally, new arrangements were discussed on the provision of international official liquidity to countries or financial markets, including the question of IMF being given the authority and the means to act as the lender of the last resort. The initiatives that were considered in this area can be summarized as follows.

It was generally agreed that the IMF should continue to play an important role in providing international liquidity. The major issues regarding adequacy of funds related to increase in quotas, SDR allocation, sale of gold to augment the resource base and increased borrowing from members. Proposals for enhancing the accountability and legitimacy of the World Bank also received some attention. As regards the proposal to establish an international lender of last resort, one line of argument was that there is no such need. A better approach to crisis management, it was argued, would be to reform the debtor–creditor relationships, including the introduction of provisions for orderly debt workouts and arrangements for temporary standstills. A number of other questions had also been raised. Who would function as the lender of last resort? Some favoured the IMF, while some others favoured a complementary facility for unconditional official lending.

## Conclusion

In conclusion, there are a couple of 'overarching' issues which deserved to be specially kept in view while considering the arrangements for financial cooperation in the new millennium. An important consideration that the international community needed to pay attention to was the hard reality that these new arrangements could not operate successfully without equal partnership between developed and developing countries, or between capital surplus and capital borrowing countries. The recent history of the time had forcefully demonstrated the close linkages between all countries across the globe, whether they were lenders or borrowers. The move to involve developing countries more closely in the discussions on the New Financial Architecture was, therefore, welcome. But initially, the institutional arrangements for decision-making on the new financial architecture still remained too heavily weighted in favour of industrial countries.

Despite many challenges, and despite the lopsided voting structure of the Bretton Woods institutions (BWIs), by and large, the consensus was that there was no urgent need to create new international institutions. At that point, the focus was on measures to make the existing BWIs more responsive and accountable. To achieve this objective, it was essential to provide greater representation to developing countries on the boards of these institutions, and to provide them with a larger voting power. It was one of the ironies of the preceding 40 years that although developing countries, as a group, had grown much faster than the developed countries over this period and their relative economic strength in terms of output and trade had increased substantially, their actual voting power in BWIs had tended to decline!

Finally, for the developing countries, it was of utmost importance that they accorded the highest priority to strengthening the banking and financial systems and bringing them up to best international standards. In early development economics, little or no attention was paid to the role of financial intermediation as an essential component of growth and development policies. The concentration at that time was on measures to raise the rate of domestic savings and investment and on the acceleration of the pace of industrialization through direct intervention. In view of the realities of that time and the historical background, this emphasis was understandable. However, the world has changed a great deal since then. The role of financial intermediaries in mobilizing and allocating domestic savings as well as external capital has become crucial. The East Asian crisis had also demonstrated the vital importance of financial institutions in sustaining the momentum of growth and development.

# 5

## EXCHANGE RATE MANAGEMENT

### 2002

In view of the considerable turbulence and volatility faced by the forex markets in several countries, policy issues relating to the management of the external sector, particularly the appropriate exchange rate systems figured very prominently in the discussions on international financial architecture in various fora, such as the IMF, the World Bank, Financial Stability Forum and the Bank of International Settlements. India had been participating in all these discussions along with central bank governors, and finance ministers of industrial and other developing countries. The preceding few years had also provided a fair amount of experience on behavioural and regulatory aspects of forex markets. Both in theory and in practice, the state of debate was still somewhat unsettled. There were a number of issues on which worldwide consensus was still evolving.

An important issue which had been extensively discussed in the literature as well as in different international fora was that of an appropriate exchange rate regime, particularly for emerging markets. There was a consensus on the so-called impossible trinity discussed earlier.

If CAC is accepted, according to accepted theory, a country either has the choice of giving up monetary independence and setting up a currency board or giving up the stable currency objective and let the exchange rate float freely so that monetary policy can then be directed to the objectives of inflation control. In this scenario, exchange rate should matter only if it affects domestic inflation. In theory, the recommended approach is either free float or a currency board.

In reality, however, the actual policy adopted by most central banks is different than the theoretical optimum. By far, the most common exchange rate regime adopted by countries, including industrial countries, was neither a currency board nor a free float. Most of the countries had adopted intermediate regimes of various types, including fixed pegs, crawling pegs and fixed rates within bands. By and large, countries had 'managed' floats or central banks intervene periodically. This was also true of the European Central Bank and the Bank of Japan traditionally. The US had also intervened with these two banks in favour of moderating the movements of the euro or yen. It was thus a matter of fact that, irrespective of the pure theoretical position of a currency board or a free float, the external value of the currency continued to be a matter of concern for most countries and central banks.

Part of the reason why countries are concerned about exchange rates is psychological, and part real. Psychological because of headline effect of a depreciating currency. For the ordinary men and women on the street and the political leaders who represent them, it becomes a matter of concern as nobody wants his or her country's economy or currency to be 'weak' or 'tumbling'. However, irrational it may be, it is a fact which has to be reckoned with by all central banks. It would be nice if there was a new terminology to describe movements in exchange rates which is less emotive and less sensational. An important reason for concern with sharp movements in exchange rates is also real, as seen in East Asia, Russia and elsewhere. The contagion effect is quick and a sharp change in the currency value can affect the real economy. Exporters may suffer if there is unanticipated sharp appreciation and debtors or other corporates may be affected badly if there is a sharp depreciation, which can also lead to bank failures and bankruptcies.

A fundamental change that had taken place was the importance given to capital flows in determining exchange rate movements as against trade deficits and economic growth, which were considered important earlier. The latter do matter, but only over a period of time. Capital flows, on the one hand, are primary determinants of exchange rate movements on a day-to-day basis. For example, the US, with the largest trade deficit in the world, had the strongest currency. Europe, with a massive trade surplus, on the other hand, had one of the weakest currencies. This result is explained by movements in capital flows, which was a relatively new phenomenon at that point. The same was happening the world over—East Asia, New Zealand, South Africa and Australia.

Capital flows in 'gross' terms which affect exchange rate can be several times higher than 'net' flows on any day and these are also much more sensitive to what everybody else is saying or doing than is the case with foreign trade or economic growth. Therefore, herding becomes unavoidable. All dealers prefer to be wrong with everyone else rather than being wrong alone! Daily risk minimization guarantees 'herd' behaviour. In this situation, as experience shows, the central banks have to intervene in some form or other, including the mightiest and not so mighty. While the degree of intervention and management varies from one central bank to another, concern about exchange rates is a fact that one has to live with. India had 'managed' floating with non-fixed rate target. Daily movements were closely watched by the RBI. India's forex markets were relatively thin, and the declared policy of the RBI was to meet temporary demand–supply imbalances that arise from time to time. For example, because of extraordinary rise in oil prices, the RBI had been meeting the oil import requirements directly as also debt service requirements. India's objective was to keep market movements orderly and ensure that there is no liquidity problem, or rumour, or panic-induced volatility.

Interestingly, in view of the actual experience about the behaviour of exchange market, there had also been a shift in the theoretical position with regard to the 'unholy' trinity of CAC, monetary independence and exchange rate stability. Some known economists seemed to favour intermediate regimes.

A related issue that has figured in the literature is that if some management of the exchange rate is required, what is it that a country should be monitoring—nominal or Real Effective Exchange Rate (REER)? From a competitive point of view and also in the medium-term perspective, it is the REER which should be monitored as it reflects changes in the external value of a currency in relation to its trading partners in real terms. However, it is no good for monitoring short-term and day-to-day movements, as 'nominal' rates are the ones which are most sensitive to capital flows and also attract the most headlines. (For example, in respect of the behaviour of dollar–euro or dollar–yen, hardly anybody talks about the real rates of exchange of these currencies!) Thus, in the short run, there is no option but to monitor the nominal rate.

Interestingly, while everyone agrees that the value of a currency should be monitored against all major currencies, all the headlines and comments by dealers are concentrated on the dollar. One hardly hears the pound hitting an 'all-time high' against the euro or yen, or vice versa. All we would hear is the euro, yen or pound against the dollar. There is certainly a reason for it as the dollar is the currency that is mostly used in trade. However, what it means is that central banks have to pay maximum attention to the US dollar whether they like it or not. This explains why ECB intervenes depending on the value of euro in terms of the dollar rather than pound or yen.

Another interesting issue is 'stability' versus 'volatility' in exchange markets. In principle, it would be desirable if exchange rates appreciate when capital flows are strong, and depreciate when they are weak. Unfortunately, in practice, this option is not always available to central banks during periods of uncertainty or turbulence because market behaviour is not symmetrical in both directions. There is a greater tendency to hold long positions in foreign currencies and hold back sales when currencies are depreciating, than the other way round. It is also a fact that corporates, investors and FIIs prefer relative stability to volatility as hedging costs less when conditions are stable.

As mentioned earlier, another issue which has figured prominently in the debate on forex management in the early 2000s was the question of appropriate policy for management of foreign exchange reserves. A few countries where monetary policy is directed towards the sole objective of inflation control, do not maintain reserves except for operational purposes. However, in the light of volatility induced by capital flows and the self-fulfilling expectations that this can generate, there is a growing consensus for emerging market countries to maintain adequate reserves. How adequacy is to be defined is still an open question. Earlier, the rule was in terms of months of imports. Later, increasingly it was felt that reserves should also be adequate to cover likely variations in capital flows.

India took into account liquidity as well as import requirements and unforeseen contingencies in the management of reserves. For this reason, the country had added substantially to its reserves in the preceding few years. Reserves were now more than adequate to meet the oil burden as well as any other likely variations in capital flows for a fairly long period. India had followed a very careful policy to reduce short-term debt, which was lower than seven or eight years ago, and also to ensure that relatively short-term deposits from NRIs were matched by foreign assets of deposit-taking banks.

There was also a consensus on the need to make available information by the central bank on reserves, including forward liabilities, as well as market operations and turnover, so that lenders and the markets are fully aware of a country's external liabilities as well as the central bank's assets and forward liabilities. The RBI was also following international practice in this regard and was regularly publishing information on all transactions as well as reserve, liquidity and forward liabilities. The above is a bird's-eye view of the dilemmas the central banks face as well as the issues that were being debated internationally at that point in time at the level of finance ministers, central bank governors and heads of international institutions. In some ways, for those who prefer certainty in information and policy-making, the prevailing state of debate may have sounded a little baffling. They must have asked themselves as to why macroeconomists and central bankers are so dumb! In their defence, one may say that the floating exchange rate system and volatility associated with capital flows were a relatively new phenomena. Until the 1990s, the policy was to have fixed exchange rates with par values or, fixed but adjustable exchange rates, in response to changes in fundamentals. Floating rates, capital volatility, massive changes in technology and integration of worldwide markets across different time zones were a relatively new phenomena in 2002. There were also some major structural changes taking place in the organization of markets, for example, as a result of the emergence of the common European currency, whose full impact was yet not evident.

Recent changes had brought tremendous benefits to the developing world, including India. Capital and technological constraints to development were fast disappearing. As the new horizons opened, developing countries were also faced with some new challenges. With appropriate policy response, India and other countries were expected to be able to take maximum advantage from new technological and other advances while minimizing risks.

# 6

# FINANCE AND DEVELOPMENT: A SHIFTING PARADIGM

## 2017

In early development economics, during the Planning era, the role of the financial system in the process of capital accumulation was relatively limited. All allocation decisions were expected to be made by the central planning authorities and not by the financial markets. To a large extent, the financial system also had a limited role in providing incentives for savings and capital accumulation, as interest rates were controlled and generally 'repressed', and household savings were pre-empted through high levels of statutory reserve and liquidity ratio.

The above-mentioned development paradigm shifted rather sharply in the 1990s. Almost all the developing countries adopted a more market-determined strategy of development. There were several factors that contributed to this change in perception. The first and foremost reason for questioning the earlier strategy was the simple fact that the actual results in terms of growth of incomes or industrial development were well below expectations. Despite substantial increase in the domestic saving rates in several countries, including India, the growth rate of incomes was relatively low. The period of relatively low growth also coincided with a period of virtually persistent and recurring BOP crises.

The change in the development paradigm also led to a change in the perception about the role of the financial system in development. It became clear that liberalization of product markets also required a well-functioning financial system for the mobilization and allocation of savings. Banks, capital markets and financial institutions were no longer seen as mere conduits for channelling savings in predetermined directions but rather as important instruments for allocating savings among alternative investment choices according to their relative efficiency.

After the 1997 East Asian crisis, the perception about the role of the financial system in development underwent a major change. It became clear that the reason behind the collapse of the real economy was the weak financial system. Hence, proper development of the financial system was no longer regarded as adjunct to the development of the real sector, but rather as essential for growth.

These developments were supported by findings in theoretical literature, which demonstrated the critical role of the financial system in the growth process. The financial liberalization literature developed in the 1970s and '80s stressed the costs of 'financial repression', particularly interest rate and exchange rate controls that restricted the growth of financial intermediation and the real rate of economic growth. These findings were buttressed by the emergence of endogenous growth literature, which emphasized the importance of financial market as a source of innovation and productivity growth. It was demonstrated that an efficient and well-functioning financial system contributed to economic growth by raising the level of saving and investment and the productivity of capital.

Over time, it was recognized that financial markets indeed had certain special characteristics. The most important of these was the large volume of transactions and the speed with which financial resources could move from one market and one instrument to another. A related characteristic was the scope for instant 'arbitrage', as between different markets and between different types of instruments. Financial transactions were highly leveraged, and the risk of failure was transferred by actual decision-makers to innocent bystanders.

Another interesting characteristic of financial markets was the role of financial intermediaries. There were segments of financial markets, such as stock markets and bond markets where savers themselves made the decision about when and where their money should be used. Markets were, however, also dominated by financial intermediaries (such as banks, provident funds, pension funds and mutual funds), which took investment decisions as well as risks on behalf of their depositors. Yet another important emerging characteristic of financial markets was the so-called 'negative' externalities associated with them. A failure in any one segment of these markets could affect all other segments of the market, including the non-financial markets.

In view of the externalities, volatility and certain other special characteristics, it was generally agreed that financial markets had to be closely monitored and supervised. It also became evident that, in view of the growing integration of worldwide financial markets, failure and vulnerability in the domestic market in a particular country could have international implications. Similarly, problems in the external markets could create difficult problems for the functioning of the domestic markets, even if the country concerned was following prudent macroeconomic policies. This close relationship between the two markets, domestic and external, raised the question of appropriate duties and responsibilities of domestic supervisory authorities and international financial institutions.

## Lessons from the Asian Crisis

Much has been said and written about the causes of the Asian crisis and its aftermath. The literature is voluminous and, in some ways, it is as impressive as the earlier literature on the 'Asian miracle' and raises the obvious question of what developing countries must learn from their successes. It is not proposed to review this literature nor to comment on what went wrong and what policies could have been handled better either before or after the crisis. The purpose here is limited and confined to recapturing some aspects of the Asian crisis that may have a bearing on the relationship between finance and development, and the lessons that countries like India need to keep in view to avoid having to go through similar devastating experiences in the future.

An important point to remember in this connection is that even relatively small mistakes in the conduct of macroeconomic or exchange-rate policies can sometimes lead to big crises. The Asian experience is certainly mixed, and the magnitude of macroeconomic and other policy failure in different East Asian countries was not the same. However, in several of them, the degree of deviation from the best practices or prudent policies was relatively small. It may be that they persisted with the defence of the pegged exchange rates for a week or two longer than was desirable, or it may be that they did not take corrective monetary or fiscal action early enough. However, the devastation and the pain that their economies went through because of these policy mistakes were sizeable and unprecedented.

Incidentally, this was also the experience of Mexico and Argentina in early 1995, when a major emerging crisis was brought under a semblance of control by a massive international rescue effort launched by the IMF, the US and the World Bank. It is no coincidence that in all these cases, in East Asia as well as in Mexico and Argentina, the proximate cause was the relatively sudden reversal of capital flows, on which these economies had become excessively dependent. It had taken a relatively long time to build a climate of confidence and for capital inflows to rise gradually. However, it took no time for this confidence to dissipate and foreign capital to disappear. It is also interesting to note that the major reversal was not only on account of foreign lenders or investors but also on account of resident holders of domestic assets who rushed to encash or convert their holdings into foreign currency.

The point is, simply, that handling capital flows is not an easy matter. While capital account liberalization and large capital movements have brought considerable growth benefits, they have also brought with them greater potential for volatility in asset prices and financial markets, including forex markets. This can cause unanticipated damage to the real economy during periods of uncertainty about the future economic or political outlook.

Adverse expectations about a country's future during periods of uncertainty can often become 'self-fulfilling'. The fact that such volatility can be aggravated by a weak financial system, leading to severe development problems, also requires to be borne in mind. The lesson from the Mexican or East Asian episode is not an argument against capital flows or capital account convertibility. It is about the careful and judicious handling of such flows and about the pace of movement towards capital account liberalization for residents. It is also about building domestic safety nets, for example, by keeping the level of liquid foreign exchange reserves high in relation to short-term external obligations.

It cannot be denied that, despite the earlier spectacular successes, the financial systems of East Asian countries were characterized by several weaknesses. Thus, banks were not subject to effective prudential regulation and supervision. Credit expansion in these countries was large and banks took untenable positions in real estate and other unproductive assets, building up, in the process, large asset-liability and currency mismatches. Banks had also built up huge off-balance-sheet liabilities, which moved on to the balance sheet once there was adversity. Cross-border inter-bank positions were also large. Non-banking financial companies contributed to the crisis as these were subject to little or no regulation.

Corporates were also highly leveraged. External debt was available at low interest rates and the fixed exchange rates in these countries offered them a false sense of complacency, encouraging them to hold large unhedged positions. External debt was high, short-term, leveraged and concentrated in the private sector. Thus, on the whole, there was an inherent vulnerability in the financial sector, and once expectation turned adverse, this vulnerability translated itself into a panic. Standards of accounting practices, financial reporting and disclosure norms were somewhat inadequate in these countries. There was lack of transparency in the operations of market participants as well as the central banks in some cases.

Events in East Asia have certainly highlighted the two-way interaction between the financial sector and development, and the need for an appropriate policy framework. Improving the efficiency of the financial sector through market-based reforms is an important concern of the development paradigm. However, this has to be accompanied by policies, practices and certain amount of restraints that strengthen the financial system towards stability, so that growth becomes sustainable. At the same time, proper emphasis has to be placed on growth policies that do not give rise to problems that result in systemic instability in the financial sector (for example, a large fiscal deficit).

A related issue is that of striking an appropriate balance between financial regulation and market freedom. While freedom is essential to foster efficiency, it also raises an equally important question of an appropriate regulatory framework, given the wide divergence between private and social interest in ensuring the stability of financial system. Hence, a proper system of regulation relating to prudent risk limits, short-term foreign borrowing and the degree of tolerable maturity mismatches in the banking system assumes critical importance for minimizing risks to the stability of the financial system.

The most important lesson emerging from the Asian crisis is the need to be vigilant about domestic and international developments that may impinge on a country's financial relations with the rest of the world. The process of integration of worldwide financial markets has resulted in product innovation and efficiency, but it has also made developing countries subject to greater vulnerability and new risks. Strong fundamentals alone cannot provide full immunity from a crisis. There is a need to take early preventive action, build firewalls and keep some safety nets handy.

## The Indian Experience

Against the backdrop of the lessons drawn from the Asian crisis, it will be useful to examine issues relating to India from the perspective of past experience, the present stage of development and policy framework for the future.

As is well-known, it was not until the Eighth Plan that the role of the financial sector and financial markets was given an explicit recognition in the development strategy. The emphasis on accelerating investment rate through state intervention in a number of key areas meant channelling credit to certain preferential sectors at subsidized interest rates, exercising public ownership control on banks and restricting their activities through policy prescriptions. Some of the typical features that got built into this system were the directed lending programme with high levels of cash reserve ratio and statutory liquidity ratio, the ceiling on deposit and lending rate, lending to priority sectors, branch licencing, and detailed regulation of banks' loan and investment portfolios.

As far as external finance is concerned, India relied primarily on bilateral and multilateral official development assistance and did not encourage private external capital inflows as a way to supplement domestic savings. The exchange rate was administered and there was extensive control over all foreign exchange transactions, which were subject to approval on a case-by-case basis. Because of pervasive exchange controls, the Indian financial system remained largely insulated from international markets. This, however, did not prevent India from suffering regular BOP crises year after year and becoming dependent on aid flows or credits from the IMF.

The financial system, as a result, faced little or no competition, either domestic or foreign, and costs and efficiency of transactions were not its primary concern. Productivity was generally poor and profitability low. The system was also subject to limited accountability. By the beginning of the 1990s, it was becoming evident that the system could not be sustained without a thorough revamping of its operations.

The BOP of 1990–91 provided the trigger point for reform in several sectors, including the financial sector. The reform initiatives in the financial sector started with the government appointing two committees: one on BOP, which went into liberalization of policies in the external sector and the second on the financial sector, which deliberated on domestic financial sector reforms. After 1992, the reform programme in the financial sector largely followed the broad approach set out by these two committees.

## Financial Reforms in the 1990s

In so far as reform in the financial sector is concerned, significant progress was made in the 1990s. There was a steady decline in the level of resource pre-emption from the banking system. Both cash reserve ratio and statutory liquidity ratio were reduced from their high levels of 15 per cent and 38.5 per cent, respectively, to 9 per cent and 25 per cent in 1991–92. Interest rates in various segments of financial markets were deregulated in a phased manner. This preceded the abolition of controls on capital issues and the freeing of the interest rate on private bonds and debentures. While the government borrowing rates were market-determined, there was a gradual phasing out of interest rate subsidies on bank loans. Wide-ranging reforms were initiated to develop and deepen the government securities market, money market, capital market and foreign exchange market. The so-called 'bank rate' was reactivated, regular short-term repos (repurchase agreements) were being conducted at a pre-announced rate, and a system of prime lending rate was introduced to provide direction to the movement of interest rates in the credit market.

In the sphere of external financial policy, while the exchange rate was market-determined, over the years, there was a progressive liberalization of foreign direct and portfolio investment, and approval procedures were considerably simplified. As a result, restrictions on inflow of capital into the economy were significantly reduced. There was also a significant liberalization of policy regarding industry's access to foreign equity and borrowing through long-term debt instruments. The banking sector was given a greater degree of freedom with regard to raising funds abroad and managing their external liability, subject to prudential guidelines. The end result of all these and other reforms was the growing integration among the various segments of financial markets, closer convergence of the Indian financial system with practices prevailing in international financial markets, and greater opportunity for investors to access both domestic and international markets.

Competitive conditions in the banking industry were facilitated by relaxing entry and exit norms and permitting the public-sector banks to raise additional capital from the market (up to a certain level). While public-sector banks continued to be predominant, the changing competitive environment in the banking sector made a substantial difference in banking practices and disclosure requirements.

Prudential regulation and supervision also formed a critical component of the financial sector reform programme. India adopted international prudential norms and practices with regard to capital adequacy, income recognition, provisioning requirement and supervision. These norms were progressively tightened over the years, particularly against the backdrop of the Asian crisis. The required capital adequacy ratio was increased to 9 per cent from 8 per cent in the banking sector. The mark-to-market practice for valuation of government securities was also gradually enhanced from 30 per cent in 1992–93 to 75 per cent by 1999–2000. As a further prudential measure against credit and market risks, risk weights were made applicable to government and other securities to take account of price variations.

An attempt was also made to avoid the problems arising from 'connected lending'. The exposure of individual banks and NBFCs to any particular borrower or groups of borrowers were prescribed, and the banking system's exposure to real estate was also limited. Prudent limits were placed on the financial system and the corporate sector on foreign borrowings.

In the area of supervision, a full-fledged institutional mechanism was developed keeping in view the needs of a strong and stable financial system. The system of off-site surveillance was combined with periodical on-site supervision for monitoring the risk profile of banks and their compliance with prudential guidelines.

As a result of these and other measures, some progress was noticeable in the performance of the Indian banking system. The trend in erosion of profit and capital base was reversed. The gross NPAs of public-sector banks (without allowing for provisions), as a percentage of total assets, also saw a decline. Most of the public-sector banks achieved the prescribed capital adequacy ratios. The improved performance also enabled most of the banks to meet their capital requirements from internal resources and the market, without excessive dependence on budgetary support.

To sum up, there was considerable progress in the broadening and strengthening of India's financial system in the last decade of the twentieth century. Some of the areas which require consideration in the future are mentioned below.

## Agenda for the Future

The agenda for the future is long. Fortunately, there has been a widespread interest and debate among experts and market participants on the various aspects of financial reform that enabled India to chart out a path best suited for it. However, there are a few areas that deserve attention, on priority, in the future. First and foremost, tighter and tougher prudential standards will, no doubt, cause some pain and impose greater responsibility on banks and other financial institutions. However, given the new international focus, and externalities and linkages involved, the regulation of the financial sector is no longer a matter of choice or a matter of domestic concern alone. Over a period of time, it is likely that the willingness of the rest of the world to do financial business—by way of trade credits, direct investments or other types of investments and loans—will depend on their confidence in India's financial practices. India must remain ahead of the curve in its prudential management.

The level of NPAs of the banking system in India had shown some improvement in the 1990s, but it was still too high. A part of the problem in resolving this issue was the carry-over of old NPAs in certain declining sectors of industry. The problem was further complicated by the fact that there were a few banks that were fundamentally weak, and their potential to return to profitability, without substantial restructuring, was low. The committee that was set up to look into the problems of weak banks made certain recommendations that were considered by the government and the RBI. These were also widely debated so that an acceptable long-term solution could be evolved. Leaving aside the problem of weak banks, in profitable banks, too, the NPA levels were found to be relatively high. In the future, vigorous effort has to be made by these banks to strengthen their internal control and risk-management system, and to set up early warning signals for timely detection and action. The resolution of the NPA problem also requires greater accountability on the part of corporates, greater disclosures in the case of defaults and an efficient credit information system. With the help of stricter accounting and prudential standards, the problem of NPAs in the future could be effectively contained.

In order to allow for growth in their assets in line with real growth in the economy, banks and financial institutions would also need to increase their capitalization quite substantially over time. The minimum shareholding by the RBI in the State Bank of India (SBI), prescribed by legislation in the 1990s, was 55 per cent. In the 1990s, the minimum percentage of shareholding by the government in public-sector banks was 51 per cent. A number of strong banks were able to access capital markets to meet their capitalization requirements in line with prudential guidelines. However, some of these banks, including the SBI, had limited scope with regard to raising further capital from the market within the prescribed floor of the RBI and government shareholding.

In this situation, an issue that needs to be resolved is whether the gap in public-sector banks in respect of capital requirements should be filled by contribution from the RBI (in the case of SBI) and the government (in the case of other public-sector banks), or whether the legislative ceiling for capital to be subscribed by the public should be raised. The provision of additional capital by the RBI is tantamount to additional monetization, and its monetary impact is equivalent to that of printing additional currency. Contribution to banks' capital by the government has a similar effect as it will add to the government's deficit, which is already high. The government, in any case, would need to provide additional capital to weak banks, which are not in a position to raise capital on their own. Does it make economic or fiscal sense to add to this burden further? On balance, there seems to be a strong case for raising the legislative ceiling for market participation in the equity capital of public-sector banks.

Over the years, the progressive liberalization of financial

markets and institutional reforms has led to growing interlinkages among the various segments of financial markets. The emergence of different types of financial intermediaries, in addition to banks and financial institutions, is healthy and desirable. A diversified structure contributes to greater stability of the financial system in the event of unanticipated problems.

In India, while there has been progress in developing the various segments of markets, including the money and debt markets, the depth of these markets remains low and the volumes as well as the number of participants is not very large. An important priority for the future is to develop the depth and breadth of these markets and to allow a multiplicity of intermediation possibilities with different risks and leverage profiles. The RBI should continue to work with financial experts and market participants to develop an appropriate procedural and policy framework to move in this direction.

India also has to devise measures to make the interest rate structure more flexible in order to take account of changes in economic cycles and the inflation outlook. For reasons highlighted in the 'Mid-term Review of Monetary and Credit Policy' in October 1999 by the RBI, there are several constraints that limit the flexibility of interest rates in the banking sector and the rest of the financial sector. Given the fact that some of these constraints are deeply embedded in historical practices, consumer preference and public-sector requirements, it may take some time to fully meet this objective. However, the process should begin.

These are just a few priority areas that require consideration. The list is by no means exhaustive. If India gets this right, it would make movement in other areas of financial reform speedier and easier.

# 7

# INDIA'S ECONOMY IN THE TWENTY-FIRST CENTURY

## 2019

In the light of several positive changes in India's comparative advantage, there is no doubt that India now has the capacity to achieve in the next 10 years what it could not achieve in the previous 50 years to eliminate poverty. India today has the knowledge and skills to produce and process a wide variety of industrial and consumer products and services. Another important factor in India's favour is international capital mobility and the integration of global financial markets. Increased mobility of capital has ensured that global resources can flow into countries that show high growth and high returns. It is possible now for India to take advantage of the virtuous circle of higher growth, higher external capital inflows and higher domestic incomes and savings, which in turn, can lead to further growth.

Increasing globalization also makes national economies vulnerable to developments outside their own borders. In addition to the direct effects, the indirect effects due to a 'contagion' can be quite serious. A crisis in any one or a group of countries can be transmitted to other countries—including those that may not have any strong economic linkages with crisis-affected countries. While the crises may initially occur in one or two specific countries, their adverse effects ripple across the world.

The lessons of these developments are clear: advances in information technology (IT), the increasing role of services, the integration of financial and capital markets and the diminished role of distance provide tremendous new opportunities for countries like India. These changes have been facilitated by unprecedented and unforeseen advances in computer and communication technology. The boundary between goods and services is also disappearing. Many industrial products are not only manufactured, but they are also researched, designed, marketed, advertised, distributed, leased and serviced. An important aspect of the 'services revolution' is that geography and levels of industrialization are no longer the primary determinants of the location of facilities for the production of services.

In future, after the impact of the COVID-19 crisis on India's economy is eliminated, the above developments provide opportunities for substantial growth. The fastest-growing segment of services is the rapid expansion of knowledge-based services, such as professional and technical. India has a tremendous advantage in the supply of such services because of a developed structure of technological and educational institutions, and lower labour costs. Another favourable development is that now it is feasible to 'unbundle' production of different types of goods and services. India does not necessarily have to be a low-cost producer of certain types of goods (for example, computers or discs) before it can become an efficient supplier of services embodied in them (for example, software or music).

At the same time, it must be recognized that the 'death of distance' and the growing integration of global products, services and financial markets in recent years have also presented new challenges for the management of the national economy—not only in India but all over the world. The trend towards integration of markets, particularly financial markets, is by no means an unmixed blessing. A heavy price may have to be paid by national economies for somnolence, sloth and non-conformity to generally accepted international norms and standards of macroeconomic management, disclosure, transparency and financial accountability.

India has to move vigorously to utilize the expanding opportunities in trade and become a location of choice for industries and services so that growth rates, along with employment, are substantially enhanced. It must also install safety nets by building a diversified and efficient financial system to aid and protect the development process at all times.

## India's Tryst with Destiny

There has been a fair amount of debate in the country on the implications of the new directions in India's economic policy. There appears to be no major disagreement about the ends or objectives of the new economic policies. Most, although not all, commentators seem to agree that efficiency in the use of resources and performance of the economy in terms of growth need to be improved. There is also a broad consensus that fiscal deficits should remain close to the projected levels, exports should be increased and more should be done for improving the health and education of the poor.

There is, however, a controversy on the instruments to be used to achieve these objectives, and whether current policies will succeed in raising the growth rate of the economy in the long run. Concern has also been expressed on certain political and social implications of the new policies, particularly whether they will make India more vulnerable to external pressures and hurt the development of the country's domestic resources. These issues are important and deserve to be fully addressed in the evolution of India's economic policy in the twenty-first century.

The first step would be to study the nature and character of state apparatus and India's previous record. India cannot merely decry the widespread corruption and 'criminalization of politics'— a phrase that has figured prominently in the Indian parliament in recent years—on the one hand, and ask for more discretionary powers for the State on the other. To do so is to suffer from a form of 'cognitive dissonance'. It is also important to remember that the results of India's old policies in promoting growth or reducing poverty were relatively unsatisfactory and worse than those registered by more open and competitive economies. As for external pressures, there is hardly any doubt that an aid-dependent economy with persistent BOP crises (as was the case in India from 1956 to 1991) is likely to be much more vulnerable than a country which is not dependent on aid or emergency assistance from abroad.

Keeping in view the actual economic and non-economic results of the old strategy, the phenomenal changes that have taken place in the world economy and India's present comparative advantage, the present direction of policies to make India more open and more competitive deserves to be accelerated. Without a radical transformation of economic policies and efforts to align them with the contemporary realities of global trade, investment and technology, it is not feasible for India to occupy the high ground and realize its full potential for growth and development.

A decisive move towards better and deeper economic reforms is the first strategic priority for the future. In developing countries such as India, handicapped with high percentages of illiteracy and poorly developed infrastructure, the government continues to have a crucial role in creating the necessary conditions for growth through investments in areas such as education, healthcare, water supply, irrigation, infrastructure, etc. These tasks cannot be taken over by the market. Successful economic reforms must result in strengthening the ability of governments to generate high growth, revenue and productivity. As the experience of several transitional and emerging market economies shows, economic reforms are necessary but by no means sufficient for growth and development.

In this connection, a core issue with multiple dimensions that needs to be resolved in the years to come is what can perhaps be best described as growing 'public–private dichotomy' in our economic life. It is a striking fact of our present situation that economic renewal and growth impulses are now occurring largely outside the public sector—at the level of private corporations (such as software companies), autonomous institutions (for example, IIMs or IITs) or by individuals at the top of their professions in India and abroad. In the governmental or public sector, on the other hand, there is marked deterioration at all levels—not only in terms of output, profits and public savings but also in the provision of vital public services in the fields of education, healthcare, water supply and transport. Most of our public resources are now dissipated in the payment of salaries or interest on past debt, with few resources available for the expansion of public or publicly supported services in vital sectors.

India can be justifiably proud of the fact that the 'rule of law' prevails in the country. Although delays in the judicial process may be legendary, there is a widespread respect for the legal system. However, for historical reasons, it is also a fact that our legal system provides full protection to the private interests of the so-called 'public servant', often at the expense of the public that he/she is supposed to serve. In addition to complete job security, any group of public servants in public-sector organization—hospitals, universities, schools, banks, bus services, post offices, railways, municipalities, etc.—can go on strike in search of higher wages, promotions and bonuses for themselves, irrespective of the costs and inconvenience to the public, in whose name they have been appointed in the first place. Problems have only worsened over time and there is little or no accountability for the public servant to perform his/her public duty.

The 'authority' of governments, both at the Centre and in the states, to enforce their decisions, has eroded over time. Governments can pass orders, for example, for the relocation of unauthorized industrial units or other structures, but implementation can be delayed if they run counter to the private interests of some (at the expense of public interest). Similarly, governments may decide to restructure public utilities to cut down waste or output losses, but these decisions are often not implemented if they adversely affect the interest of public servants employed in these organizations.

Over time, the processes and procedures for conducting business in government and public-sector organizations have become non-functional. There is a multiplicity of departments involved in the simplest of decisions, and administrative rules generally focus on the process rather than results. There is very little decentralization of decision-making powers, particularly financial powers. Thus, while local bodies have been given significant control in some states for implementing national programmes, their financial authority is limited.

The multiplicity of functions and responsibilities placed upon ill-equipped and ill-trained staff in public offices and local institutions make it almost impossible to deliver services with any degree of efficiency, particularly in rural areas. For example, a 'multipurpose' female health worker is required to perform as many as 40 or more tasks on a regular basis.

People who suffer the most from fiscal stringency and the atrophy of public delivery systems are certainly not the affluent or those elected or appointed by the public to serve them. They can always go to private hospitals, private schools, autonomous universities or institutes of higher education, etc., to meet their requirements. The worst affected are likely to be the poor, the unemployed and the illiterate, who rely on public services, public investment and public programmes. It is useful to keep this perspective in mind, since a fair amount of argument in favour of maintaining the status quo of the public sector is made in the name of the weaker sections of the society.

In addition to economic reforms, which already figure prominently in the national discourse, it is now important to embark on an urgent programme to revitalize the governance and public delivery systems at all levels of governments—the Centre, state and districts. Without strengthening the ability of the government to do what it alone can do and narrowing the focus of its activities to what matters most for the future development of the country—education, healthcare, a clean environment and a functioning infrastructure—India cannot adequately seize the opportunities that lie ahead.

While there will be problems and arguments for and against a particular policy to liberalize, open up or introduce more competition, actual progress in the desired direction is unavoidable and irreversible. At the same time, it is also clear that unless India improves its fiscal expenditure pattern and public delivery systems, no amount of macroeconomic policy reforms by themselves will be sustainable or yield permanent results. If lakhs of primary schools run by government agencies, thousands of primary healthcare centres set up by district authorities and hundreds of central and state universities continue to underperform or decline, a few hi-tech cities, new business schools and technical institutes cannot make up for the gigantic waste of human resources. It is simply a question of relative proportions and inextricable linkages between public good and private progress.

To overcome some of these problems, India needs to move on a number of fronts. Legal reforms should focus sharply on the interests of the public and not those of public servants. Clear mechanisms for establishing accountability for performance are essential, and all forms of special protection for persons working for government or public-sector agencies (except for the armed forces or agencies engaged in the maintenance of law and order) deserve to be eliminated. All public monopolies should be removed, and there should be no purchase preference for public-sector enterprises or agencies. The government should be free to engage the services of non-governmental organizations (NGOs) or private service providers at competitive costs to ensure effective delivery of essential services. There should be full disclosure of all financial decisions made by the government and its multifarious agencies on a daily rather than quarterly or annual basis. India needs a new 'political-bureaucratic' compact based on a well-defined division of responsibility and accountability. The public system cannot function without respect for conventions and mutual trust and harmony among different agencies of the state. Above all, there is an urgent need for fiscal empowerment at the state and local level. The last is the most difficult task in view of the deadweight of the past, but it can no longer be avoided.

If, with its high economic potential and a majority-party government in power, India embarks on a programme to revitalize its governance and public delivery systems at all levels of government, there is simply no doubt that it can realize its full potential to eliminate poverty and emerge as one of the fastest-growing economies of the twenty-first century.

## SECTION II

# BEYOND THE METRICS OF ECONOMY

# 8

# THE ROLE OF PARLIAMENT

## 2007

The Parliament of India is truly representative of the vast economic, social, regional and religious diversity of India. All income classes—from the richest industrialist to the poorest farmer—are represented. All castes and all regions find equitable representation depending on their size, population and electoral popularity. Members belong to different religions and can openly and freely espouse their beliefs, irrespective of their numbers. In the midst of this great diversity, there is also unity. Every member has a single vote and an equal right to intervene in the debate independently or on behalf of a party. The time and space allotted to party or non-party members is also equitably distributed depending on their numbers. Ministers speak on behalf of the government, but have no special privileges or ostentatious perquisites or attendants inside the House. Any member is free to interrupt, shout or otherwise disrupt the proceedings of the House, irrespective of seniority, and is subject only to the directions of the Chair inside the House. While there is discussion and debate on important matters, and there are strong political differences among the parties within and outside the government, most legislative proposals and official resolutions are adopted without dissent.

The parliament is the supreme forum of India's democracy, and represents the will of the people and their different identities.

Except for some brief aberrations (such as, during the period of Emergency, 1975–77), successive governments have also been sensitive to the views of parliament on issues of high national policy, foreign affairs and defence.

While all this is true, over the years, there has been a subtle change in the role of parliament below the surface, which is not evident at first glance. All citizens who follow the news in the media or who watch parliamentary proceedings are aware of, and perhaps disappointed by, the frequent disruptions that now occur in the two Houses. The concern with the functioning of India's parliament and state legislatures was also voiced by the National Commission to Review the Working of the Constitution. The Commission highlighted the fact that even the relatively fewer days on which the Houses meet are often marked by unseemly incidents, including the use of force to intimidate opponents, shouting and shutting out of debate and discussion resulting in frequent adjournments. There is an increasing concern about the decline of parliament's falling standards of debate, erosion of the moral authority and prestige of the supreme tribune of the people.

In the context of coalition politics, there is also increasing acceptance by political leaders of the frequent violation of democratic norms and conventions in the political decision-making process. As a result, there is a possible threat to the preservation of the cherished goals of 'Unity in Diversity', which is an important gift to the nation from leaders such as Mahatma Gandhi and Jawaharlal Nehru, in the early years of independent India. Some signs of the increasing divide in the national mainstream are already evident. In 2007, as many as 160 districts of India were under the influence of Naxalites and function largely outside the control of state governments.

To what extent the growing power of militant movements reflects the weakness of the State is a moot question. It is a fact that in several states, where lawlessness has spread in a large number of districts, the administration has been extremely weak. Political leadership has been ineffective and there have been frequent and arbitrary transfers of senior police officers and other district officials. The duality of India is also evident in the increasing income disparities among the people, seen in the vast contrast between India's rising global economic clout, as reflected in the large number of Indians in the list of world's billionaires, and the deteriorating conditions in its rural areas, where more than 70 per cent of its citizens live. This divide is also reflected in the divisiveness at the highest levels of the government, where ministers and leaders belonging to different parties are inclined to follow their own agenda rather than a collective and shared vision for the nation's future.

In what follows, some recent instances are highlighted where the proceedings of parliament, including its silences, posed serious challenges to the functioning of India's democracy as a unifying force among people with a diversity of interests, identities and outlook. Incidentally, what is said below largely reflects an eyewitness account of happenings in the Upper House of the Indian parliament, the Rajya Sabha. In many ways, the Rajya Sabha has also lost its separate identity, as what happens in this House largely reflects the positions taken by the different parties in the Lok Sabha, the House of the People. If the Lok Sabha is disrupted, so is the Rajya Sabha, and vice versa. If a bill is passed in the Lok Sabha without discussion because of disruptions or because sufficient time is not available for discussion, the Rajya Sabha is also likely to follow suit.

## Taxation without Representation

In the annals of India's long and distinguished parliamentary history, the events that took place over five days, between 18 March and 23 March, during the Budget session of 2006 were perhaps unique. Over the course of these five days, a number of unexpected decisions were announced by the government regarding the business agenda of the two Houses, which were passively accepted by both the Houses. These decisions involved a major change in the established procedure for consideration of the Budget, a drastic revision in the business of the two Houses without adequate notice and a sudden adjournment of parliament *sine die* (followed by a reversal of this decision again a few days later). The passive and ready acceptance by parliament, the supreme institution of India's democracy, of decisions that are contrary to well-established parliamentary conventions has serious implications for the future. It is, therefore, worth going into the events of these five days in some detail.

As per the usual procedure, the Budget session of parliament for 2006 was convened by the president to meet in two parts: from 16 February to 17 March and again from 3 April to 28 April. However, on 7 March 2006, in view of the elections announced by the Election Commission of India (ECI) in five states over the months of April and May, it was decided to have a longer interval between the two parts of the Budget session. The dates announced earlier for the two parts of the session were changed, and it was decided to hold the first session from 16 February to 22 March, and the second session from 10 May to 23 May. The first part was longer and the second part was a bit shorter than the original schedule, but on the whole, the entire Budget session was supposed to be long enough to permit the examination of the Budget as per established convention.

It will be recalled that, according to Rules 272 and 331G of the Rules of Procedure and Conduct of Business in the Rajya Sabha and the Lok Sabha respectively, it is mandatory for the demands for grants of the ministries and departments of the Government of India to be examined by the concerned standing committees of parliament (which were set up in 1993). The standing committees consist of members of both Houses of parliament. The agenda and the meetings of the committees are conducted by a chairperson, who is normally a senior member of one of the Houses. The examination of the Budget grants by these committees allows members, belonging to both Houses and to different parties, to question the senior representatives of the ministries or departments, and also to hear and examine other witnesses, including members of NGOs and experts. The observations and recommendations of these committees are normally unanimous and non-partisan. The reports of these committees on matters under their purview, including the Budget demands, are submitted to the two Houses of parliament for consideration.

In order to allow the standing committees sufficient time for careful consideration of the Budget demands, it has also been the convention for the Houses of parliament to adjourn for about a fortnight between the two parts of the Budget session. The first part of the session is devoted to a general discussion of the Budget by members and for the reply by the finance minister on broader macroeconomic aspects. The reports of the standing committees on the ministries/departments are considered in the second part of the session followed by voting on demand for grants and consideration of the Finance Bill for the new fiscal year.

In 2006, as it happened, after the changes in the dates of the two parts of the Budget session were announced on 7 March, a controversy arose over the definition of the so-called 'office of profit'. Some members were alleged to have been appointed to such offices by state and central governments, which is not permissible under the Constitution. One well-known member was also disqualified as Member of Parliament (MP) on these grounds by the president on the advice of the ECI. It was in the context of this controversy that a number of decisions were announced by the government, and were accepted by parliament, which violated several well-established conventions and norms.

Thus, on 18 March 2006, all of a sudden the government decided to introduce a motion in the Rajya Sabha for the suspension of Rule 272 (and for a similar motion for the suspension of the relevant rule in the Lok Sabha). The motion to suspend consideration of Budget demands by the standing committees was moved and adopted without discussion in the two Houses on the same day. With the suspension of consideration by standing committees, the ground was cleared for the adoption of the Budget as well as the Finance Bill in the first part of the session itself. This was an extraordinary and unprecedented event in a year when there was no change in government, no general election and no internal or external emergency. And yet it was decided to rush the Budget through parliament without proper consideration.

Rule 272 was suspended in the Rajya Sabha on 18 March and Rule 331G was suspended in the Lok Sabha on the previous day. There was a session of parliament on 19 March, which was a Sunday. On Monday, 20 March, the consideration of the Budget demand for grants (or the appropriation bill), as passed by the Lok Sabha on 18 March, was listed in the revised list of business in the Rajya Sabha. The controversy over the 'office of profit' issue became more intense because of allegations and counter allegations by major parties regarding top party leaders holding various offices of profit under the central and state governments and still continuing as MPs. Nevertheless, the budget appropriations were considered and approved by the House on the same day. On the next day (i.e., Tuesday, 21 March), the Finance Bill, i.e. the bill to change tax laws, was listed in the revised list of business and was duly approved by a voice vote in the midst of considerable noise and disruption.

Developments in parliament on Wednesday, 22 March, were, however, even more extraordinary and unexpected—and in some sense, bizarre. Before parliament met in the morning on that day, there was a strong rumour that the ruling coalition was considering exempting certain offices from the purview of the proposed offices of profit legislation by issuing an ordinance after the first part of the Budget session ended in the evening. The reason for this extraordinary move, as reported in the press, was to ensure the continuation of the Congress president in parliament. She was also holding the office of chairperson of the National Advisory Council (NAC) with Cabinet rank. Unfortunately, at the time of her appointment, the government had not taken steps to exempt this office, which could have been done easily and without any controversy.

The opposition parties, as a mark of protest, decided to disrupt the parliament on 22 March, and not allow any listed business to be considered (the Union Budget had already been passed on the previous day). After an obituary reference, which lasted for about four minutes when the House met at 11 a.m., in view of shouting by some members, it was decided by the chairman of the Rajya Sabha to adjourn the House for 20 minutes (from 11.10 a.m. to 11.30 a.m.). The House met as scheduled, but was again adjourned after four minutes of disruption, and was asked to meet at 1.00 p.m. However, during those four minutes, more than a hundred papers, including annual reports of public-sector organizations, outcome and performance budgets, reports on action taken, and the notifications issued by various departments of the government were laid on the table of the House by a dozen ministers in the midst of pandemonium. After assembling at 1.00 p.m., the House had to be adjourned for the third time without conducting any business. It was asked to re-assemble at 2.00 p.m.

The House met for the fourth time that day at 2.00 p.m. and was adjourned after two minutes for half-an-hour. Again, there was a disruption and it was adjourned until 5.00 p.m. The House met for the sixth and last time at 5.00 p.m. This last session, which lasted for only 15 minutes, completed all the listed business for the day, including the adoption of a legislative bill without any discussion, in a noisy and disruptive House.

No explanation was given in the House for the reasons behind deciding to suspend the Budget session after the first part. However, in response to questions by the media, it was explained by the government that the House had been adjourned *sine die* because the Budget had already been passed and hence there was hardly any business left to be transacted.

The end of the Budget session on 22 March was followed by a surprise announcement the next day, 23 March, by the Congress president. She decided to renounce her seat in parliament, and seek re-election after resigning from all other government positions (including that of the chairperson of the NAC). According to media reports, in the light of this unexpected development, the government had no option but to give up its plans to issue an ordinance exempting certain offices from the purview of 'office of profit' rules. In subsequent press interview, it was announced by the concerned minister that the government would consult other parties in parliament and bring about appropriate legislation for consideration in due course.

After four days of abrupt *sine die* adjournments, the government announced its intention of reconvening parliament, as earlier scheduled, from 10 May to 23 May 2006. A formal notice to this effect was also issued to all members on 5 April 2006 after the necessary formalities had been completed. On 28 March 2006, members were also informed that notwithstanding the completion of discussion and voting on the Demands for Grants of the respective ministries/departments for the year 2006–07 by the Lok Sabha, it has been decided that the department-related parliamentary standing committees will examine these Demands and Grants and present their reports thereon to the respective Houses.

Thus, the standing committees were also resurrected as suddenly as they had been dispensed with—even though there was nothing left for them to consider, recommend or approve. This move was yet another step in the direction of diminishing the role of parliament in the conduct of the nation's affairs.

The parliament now does what the executive decides or does not decide, presumably after some behind-the-scene consultations with selected party leaders. The events of 18–22 March, and the subsequent decision to reverse some of the unconventional decisions taken earlier, are perhaps a culmination of a process marked by ad hocism and expediency in the functioning of parliament. There are two other examples of the shrinking role that parliament now performs. These examples, taken from two previous sessions of parliament, are perhaps indicative of an emerging trend and the direction in which parliament's role is now drifting. On 29 August 2005, a day before the end of the monsoon session in the previous year, the Rajya Sabha adopted an important bill, the Women's Succession Bill, in four minutes flat, in the midst of shouting and the shutting out of any debate and discussion on the bill. Members were not even able to hear the minister rising to introduce the bill. The clause-by-clause consideration was also taken up without any member being able to speak or comment in the midst of a disruptive and noisy House. Then the bill was passed by a voice vote, with most members not even being aware that the Chair had asked for such a vote. Fortunately, this particular bill was concerned with providing equal treatment of all citizens, in respect of inheritance and gender, and it enjoyed wide public support. However, what was alarming was not the contents of the bill, but the way in which it was passed. Based on this precedent, at least in principle, any other bill, whatever its contents, is capable of being passed in the same way.

Similarly, in an uncanny resemblance to the procedure adopted for the passage of the Budget for 2006, on 26 August 2004, parliament had also decided to suspend the question hour and pass the regular Budget involving an expenditure of more than ₹4,75,000 crore without any discussion, within a few minutes. This was also the result of a backroom agreement between the leaders of the parties in the government and the Opposition, following several days of disruption of parliamentary work (because of a dispute on a sensitive but extraneous matter).

In 2004 and 2005, when the above events occurred, they had seemed unusual and somewhat alarming. However, in the light of developments that took place during the five days of March 2006, they pale into relative insignificance. During those five days, not only was the Budget passed abruptly in advance of the normal schedule, but an important bill was also adopted without discussion or advance notice in a disrupted House. And then the session was adjourned *sine die*, only to be reconvened again!

It may be argued that the primary responsibility for the above series of events lies with a disgruntled Opposition, and not with the government. It was the Opposition that was indulging in frequent disruptions in the two Houses and the government had no option but to somehow carry on with the task of running the affairs of the nation. This contention may have some validity, but it does not resolve the issue of the complete subservience of parliament to the will of the executive. If bills can be passed, if Budgets can be approved, and if sessions can be adjourned abruptly, an irresponsible or autocratic government can easily get away with the erosion, and even the suspension, of the legitimate rights of the people. So far, with one or two possible exceptions, the country has been fortunate in having been led by leaders of integrity and democratic values. However, there is no guarantee that this will continue to be so in the future.

There is also no legitimate explanation for the decision to end the Budget session well in advance of the announced schedule or to suspend the procedure for the examination of the Budget by the standing committees and then reverse these decisions arbitrarily after a couple of days. The sanctity of well-established convention and practices deserves to be preserved rather than abandoned on grounds of expediency. This is feasible if parliament, rather than the executive in power, is in charge of its own functioning and if the Chairs of the two Houses are given adequate powers to control an unruly Opposition. It is the duty of a democratic and elected government, not only to somehow carry on with the business of governance, but also to ensure that the means adopted for doing so conform to democratic best practices and to the intent of the

Constitution to make the executive accountable to the legislature rather than the other way round.

## The Silences of Parliament

In addition to approving legislative proposals and other government business, the parliament is an important forum for the discussion of public issues and public grievances through their representatives. There are regular question hours for members to ask questions of their choice concerning different ministries. Ministers are responsible for answering these questions and for taking further action as necessary in the light of discussions on 'starred' questions. Time is allotted for members to make 'special mentions' on an issue of importance to their constituents, their states and the country. A member is entitled to propose a 'short-duration' discussion on any matter of public importance. He/she can also move a resolution or a private member's bill for discussion and approval after completing the necessary formalities for doing so.

Debate on important policy issues is exhaustive, penetrating and highly useful (for example, on subjects such as the nuclear cooperation agreement with the US, the Rural Employment Guarantee Act, development problems in the least developed parts of the country and enhancing regional cooperation in South-east Asia). The issues raised during the debates also influence the course of policy formulation by the government of the day. This is an important strength of India's democracy as national policies of long-term domestic and international importance, including economic policies, are adopted after careful consideration and broad consensus across the political spectrum. This explains why national policies, once approved by parliament after discussion, are seldom reversed despite changes of government.

However, there have been occasions when the silences of parliament have been just as loud as the debates on foreign policy, employment and development policy. Generally, the tolerance for deviation from established norms and propriety is most evident when the interests of a supreme leader of the party in power are under threat or when there is a clash of interests among different parties in search of political power after elections (or an adverse judicial verdict). The most conspicuous example of such silences was, of course, during the period of Emergency in 1975–77, when violations of established laws and administrative norms were either tolerated or approved through legislative amendments, including Constitutional amendments.

Fortunately, for India's democracy, such occasions have been relatively infrequent. The power of parliament to alter the fundamental rights of the people and the 'basic structure' of the Constitution has also been declared invalid by the Supreme Court of India as early as in 1973 (during the hearing on the famous *Kesvananda* case). It will be recalled that the verdict of the Supreme Court in this case was challenged in 1975 by the government after the imposition of Emergency. It was argued that parliament was 'supreme' and represented the sovereign will of the people. As such, if the people's representatives in parliament decided to change a particular law to curb individual freedom or limit the scope of judicial scrutiny, the judiciary had no right to question whether it was Constitutional or not. After listening to the persuasive arguments of legal luminaries, the Chief Justice of India decided to dissolve the Bench, and the 'basic structure' doctrine was reaffirmed as an unalienable feature of our Constitution.

The 'basic structure' doctrine has not been challenged or compromised by any party or parties in power after 1975. However, in later years, the silences of parliament have become more frequent on several issues of public interest. New state governments have been sworn in even though they did not have a majority in legislatures. Ordinances have been issued by governments without adequate cause, and prosecution of criminal offenders has been deferred to protect the political interests of some parties or powerful leaders. On such issues of paramount national importance, the parliament has maintained silence or has given its approval *post facto* under the Constitution in case such approval was required (for example, for imposition of President's Rule in Bihar in 2005 by issuing ordinance, which was later found to be un-Constitutional by the Supreme Court).

Again, fortunately for India, these cases have been exceptional, and despite the silences and tolerance of parliament, the wrong decisions taken by Constitutional authorities have generally been reversed later after judicial scrutiny. However, some unhealthy precedents have been set and it cannot be taken for granted that these will not be repeated in the future. It is useful to remind ourselves of some of the recent cases where parliament did not play its part in holding the executive accountable for its actions. The aggrieved persons or parties needed to approach the judiciary for redressal of their grievances.

In this connection, the developments that took place in the state of Bihar after regular state elections were completed in February 2005, are of particular interest. It will be recalled that the electoral verdict in this case was fractured and that no party or combination of parties had a clear majority. This included the ruling Rashtriya Janata Dal (RJD), which had been in power for several years. After considering various options, the then governor of the state was pleased to recommend the imposition of President's Rule without dissolving the assembly. However, after patiently waiting for three months, all of a sudden and without any notice or discussion with the various political parties, on 23 May 2005, he felt compelled to recommend that the assembly should be dissolved immediately. The Union Cabinet considered it appropriate to meet late at night and advise the president, who was on a state visit to Moscow, to approve the governor's recommendation during the course of the night itself.

The reason for this great urgency three months after the election was not made clear, although it was claimed by the governor later that this decision had become unavoidable because 'horse-trading' among legislators was taking place. The governor and the Government of India, therefore, considered it necessary to dissolve the assembly to prevent unethical behaviour on the part of the legislators. According to media reports and other available evidence, the real reason for the hasty action was that legislators belonging to some minority parties had decided, after waiting for three months, to join a coalition of other parties that were opposed to the RJD. As it happened, the RJD was a member of the ruling coalition at the Centre with a number of ministers in the Central Cabinet. The Centre, therefore, had no option but to take the midnight decision to prevent another coalition of parties from taking office in Bihar.

The opposition parties in the state were naturally upset by the Centre's decision, and some of the affected legislators decided to file a case against the decision of the Government of India. In defence of its case, an affidavit was filed by the central government in the Supreme Court. In its affidavit, the government argued that 'the Court is not to inquire—it is not concerned with whether any advice was tendered by any minister or Council of Ministers to the President, and if so, what was that advice. That is a matter between the President and his Council of Ministers.' In other words, according to the government, the Council of Ministers could advise the president to pass any order (irrespective of its merits); the president had no option but to accept that advice under the Constitution; and the Supreme Court had no right to examine whether the action of the executive was legal or not!

After hearing the arguments, in October 2005, the Supreme Court gave a summary verdict declaring the action of the government to dissolve the Bihar assembly as 'un-Constitutional' and unreasonable. The Court, however, did not order the revival of the old assembly as fresh elections had already been announced by the ECI and were scheduled to take place after a few days. The Court's verdict caused considerable public embarrassment to the government since the decision to dissolve the assembly had been taken by the president at very short notice on the advice of the Union Cabinet. In the light of the Supreme Court verdict, the governor of Bihar tendered his resignation. And that was the end of the matter so far as the government was concerned.

When the above events were taking place, parliament was in recess. The monsoon session of parliament was reconvened in the last week of July 2005. As per the provisions of the Constitution, the ordinance to dissolve the state assembly had to be formally approved by parliament. The government moved a bill to that effect, which was duly approved. Interestingly, after the Supreme Court verdict in October 2005, declaring that the action to dissolve the state assembly was un-Constitutional, some parties in parliament put forward the view that the Supreme Court in its judgment had exceeded its brief since the ordinance had already been approved by the Parliament of India! Subsequently, in November 2005, after fresh elections were held, a new coalition government was formed by the parties that had earlier been denied the right to test their majority on the floor of the assembly.

An even more blatant example of transgression of well-established Constitutional conventions by the governor of a state had occurred in March 2005 in the state of Jharkhand. After the elections, in Jharkhand also, there was no clear majority among the pre-election allies. However, the opposition parties were able to persuade some other elected members to join them. They were thus able to demonstrate their majority to the governor (with 41 members in a House of 80 members). However, in his wisdom, the governor decided to swear in a government headed by a member of the Union Cabinet, who did not seem to have a clear majority. He was also given a number of days to prove his majority on the floor of the House. The opposition parties that claimed to have a majority were extremely upset by this decision of the governor and filed a writ petition in the Supreme Court challenging the decision. On March 9 2005, the Court passed an order which *inter alia* gave directions to the Speaker to extend the state assembly session by a day and conduct a floor test between the contending political alliances. In the light of the Supreme Court's decision, the earlier government formed by the Union minister decided to tender its resignation on the advice of the central government. An alternative government was then formed by a combination of other parties which was able to prove its majority on the floor of the House.

The directions of the Supreme Court to the Speaker of the Jharkhand assembly raised a legal storm, as these were interpreted by several experts as intruding into an area that was within the jurisdiction of the legislature. This view was also endorsed by an Emergent Conference of the presiding officers of Legislative Bodies of India, which was convened at short notice on 20 March 2005, to deliberate on the Constitutional issue arising from the verdict of the Supreme Court. The presiding officers expressed their concern in no uncertain terms over such orders passed by the courts repeatedly which tend to disturb the delicate balance of power between the judiciary and the legislature and appear to be a transgression into the independence of the parliamentary system of our country.

In parliament, there was no disapproval of the undemocratic actions of the governor. The concern expressed by presiding officers of the legislative bodies was not about the actions of the governor. It was about the Supreme Court transgressing its jurisdiction in giving directions to the legislature for impartially carrying out the Constitutional provisions in respect of the formation of the government.

The role of state legislatures in defending the provision of the Constitution, including the procedure for the approval of state Budgets, has become even more perfunctory than that of parliament. In some states, the Budget sessions are now held for a few days only, and Budgets are passed practically without any discussion. The same is the case in regard to the approval of new laws or legislative amendments proposed by the government. Part of the reason for this state of affairs is the unbridled power of the Opposition to disrupt the House. The greater the chaos generated when the House is in session, the greater the publicity. Such publicity is considered to be a major gain for parties, particularly small parties, outside the ruling coalition.

In order to prevent the destabilization of a government by splitting a party which is a part of the ruling coalition and to prevent cross-voting during Rajya Sabha elections, two important legislative changes were adopted by parliament in April, 2003. The first amendment was that any elected member (or a group of members) who decided to leave his or her party would have to seek fresh election. The second amendment (pertaining to election to the Rajya Sabha) replaced secret voting by an open-voting process by members of legislatures. This amendment was designed to prevent cross-voting so that members who did not vote for their party's candidates could be removed from their party for 'indiscipline'. The domicile requirement of candidates for election to the Rajya Sabha was also removed.

On the face of it, these amendments seem sensible because they are designed to reduce instability and corruption among the members of a party. However, in reality, the effect has been to strengthen the powers of party leaders over their members. The solution adopted, with multi-party consensus, is in fact a lot worse than the disease. While members have no right to defect, the leader of a party is free to create instability by forcing all members of the party to leave a multi-party government, even if the majority of the members do not agree with that decision. Similarly, nomination to the Rajya Sabha has become the sole prerogative of the leader of a party (and a few persons who enjoy his/her confidence). Bribery or the funding of parties in exchange for nomination to the Rajya Sabha has also not been curbed.

While parliament sessions are held frequently and vast quantities of papers containing information on the working of ministries are placed before it, the events of March 2006 have established beyond reasonable doubt that parliament has practically no role in holding the government accountable for its performance. It is now a regular practice for government business or legislative proposals that require parliamentary approval to be approved without much debate and within a few minutes towards the end of the day, when only a few members, including those from the parties in power, are present.

There was a time when assurances given by ministers on the floor of parliament had a ring of credibility to them. Unlike other commitments, those made in the two Houses were supposed to be translated into reality if only for the fear of attracting motions of breach of privilege. This is no longer the case. Assurances in parliament are now just like any other assurance, meant to be bypassed or forgotten without explanation. Hundreds of assurances, some of them made more than a decade ago, are still pending.

# 9

# THE EXECUTIVE AND THE JUDICIARY

## 2007

As highlighted in the previous article, the responsibility of the Council of Ministers to the House of the People or the Parliament of India is largely *pro forma*. As the events of the five days in March 2006, from 18 March to 22 March, abundantly demonstrated, as long as a government can command the support of a majority of members belonging to one or more parties, it is the will of the government that prevails in parliament rather than the other way round. Parliament may be convened to meet during a specified period in the future, but it may be abruptly adjourned *sine die* without any explanation if the government so decides; the decision to adjourn *sine die* may also be reversed a few days later after parliament has actually adjourned; the Budget and the Finance Bill may be approved by voice votes at a day's notice; and a legislative bill may be passed in the midst of a disrupted House without any discussion.

Fortunately, the judiciary continues to be the final arbiter of the legality or otherwise of decisions taken by the executive either on its own or with the approval of parliament, as required. In the light of the Supreme Court's 1973 decision (in the famous *Kesvananda* case), which confirmed that the 'basic structure' of the Constitution was sacrosanct, the judiciary continues to have the ultimate power to interpret the Constitution and its intent.

While there is common agreement among all the organs

of the state, i.e., the legislature, the executive and the judiciary, as well as on their respective powers and jurisdictions, from time to time, a dispute arises about whether the judiciary has exceeded its powers in issuing directions to the other branches of the government in the light of its own interpretation of the provisions of the Constitution. The dispute about the judiciary having exceeded its powers becomes intense when the executive branch at the Centre or in the states takes a decision that is apparently in favour of some parties at the expense of other parties. Irrespective of the intrinsic merits or otherwise of the decisions taken in such a situation, an appeal to the courts by aggrieved parties and legislators becomes unavoidable.

This is precisely what happened after the electoral verdicts in three states, i.e. Goa, Jharkhand and Bihar, in early 2005. Neither the governors of these states (who had the final powers to appoint a government) nor the presiding officers of the legislatures (who had the power to conduct the proceedings of the House to test the majority) were considered to be impartial in their decisions. The aggrieved legislators went to the Court, which issued certain interim directions to the executive and/or legislature as per its interpretation of the Constitution. The directions of the Court, in turn, led to protests by members of the new government as well as the presiding officers about the Court having exceeded its jurisdiction in issuing such directions.

In India's democratic history, the political pressure to limit the powers of the judiciary and declare parliament as being 'supreme' and representative of the will of the 'people of India' is the strongest when a coalition government of parties, with varying agendas, is in power at the Centre or in states, and the political survival of the undisputed leader of the majority party is threatened (as, for example, was the case in 1975, when Emergency was imposed). In these circumstances, political survival becomes more important than the legal merits or demerits of a case.

The erosion in the Constitutional principle of collective

responsibility, the politicization of the bureaucracy, and the dispute about the separation of powers between the judiciary and the executive (or legislature) have important implications for the working of India's political system and its institutions. Some of these implications are discussed below.

## The Principle of Collective Responsibility

In the parliamentary system of government, unlike the presidential system, all members of the Cabinet are members of the legislature (the parliament at the federal level and the state assemblies at the state level). The prime minister (PM) is elected to head the government by the party that has won majority (or is selected by consensus among the parties in a coalition), and is supposed to be 'first among equals'. The PM, in turn, selects the members of the Cabinet and assigns them to different ministries and departments of the government as ministers or ministers of state. The PM is also free to create new ministries and departments or to merge and change the items of business assigned to different ministries. In the formation of new governments, depending on their personal standing in the party and the extent of political power they enjoy, PMs are free to act on their own or they may need to consult the leader of their party and its coalition partners.

Once the decisions concerning the composition of the Council of Ministers have been made and communicated to the president, these decisions are final. The Cabinet is then supposed to be collectively responsible to parliament or the legislatures. All policy decisions taken by individual ministries, irrespective of who leads them, and all laws or amendments to existing laws have to be approved by the Cabinet as a whole before they are introduced in parliament. Similarly, all important administrative decisions, including appointments, are supposed to be put up to the Cabinet or to a designated committee of the Cabinet after appropriate inter-ministerial consultations. All Cabinet decisions, once approved, are unanimous and the Cabinet is collectively responsible for them.

The principle of collective responsibility of the Cabinet has some important Constitutional implications for the conduct of individual ministers. First, no major policy or administrative decision should be taken without appropriate inter-ministerial consultations and the approval of the PM. Second, all such decisions must represent a consensus among members of the Cabinet or its committees, whether they are directly concerned with the subject under consideration or not. Third, all members of the Cabinet are jointly and collectively responsible for the performance of the government in parliament (and thus indirectly to the people), irrespective of the particular ministry to which a particular item of business has been allotted.

An important corollary of the principle of collective responsibility is that no individual minister can be formally held accountable for the failure of a ministry to implement a decision or a programme announced by the government. Thus, to take an extreme example, the government can declare war or sign a peace or border treaty, which may later be considered to be unwarranted or poorly implemented. An individual minister, however, cannot be held responsible for the wrong decision or failure in implementation. The Council of Ministers as a whole— irrespective of any internal dissention and disagreement—would have to rise in defence of the minister. The PM or the leader of the party is, of course, free to ask the minister to resign or remove him/her. However, so far as the public or parliament is concerned, he/she has no formal individual responsibility and accountability for implementing the decisions taken on behalf of the government. An individual minister cannot be suspended by parliament for any policy or administrative decision taken by his/her ministry with the approval of the Cabinet. Questions may be asked, calling attention motions may be moved and even cases may be filed in courts for impropriety or corruption, but the person can continue in office as long as the PM wishes him to and the party in power enjoys a majority in the House.

Against the above background, it is not surprising that successive governments and ministers, since Independence, have announced grand plans for removing poverty, achieving full employment and providing essential services to the people. And yet India, despite all the recent growth and shine, remains one of the poorest countries in the world, with the highest number of poor persons, who enjoy the right to vote and elect their government. Programmes aimed at removing poverty and providing services to the poor have also been the principal items on the economic agenda of every political party at the time of elections. The instrumentalities and specific policies proposed to be adopted, if voted to power, have varied, but the anti-poverty objective has been the same. The poor, of course, continue to have the power to vote, and enjoy substantial electoral power in a majority of constituencies in the country. But once the elections are over, accountability for the performance of ministers is conspicuous by its absence.

While the principle of collective responsibility has effectively shielded individual ministers from being held accountable for performance of their ministries, this principle has not prevented them from taking decisions on matters of great public importance without seeking formal approval of the Cabinet. There have been several such cases in the past. In order to appreciate the full implications of the erosion of the principle of collective responsibility, which shields ministers from taking individual responsibility, two cases are discussed here.

In March 2004, when the National Democratic Alliance (NDA) coalition was in power, the then minister in charge of higher education announced the decision of his ministry to drastically reduce the financial autonomy enjoyed by the Indian Institutes of Management (IIMs). This decision, which was taken by the minister without any reference or endorsement of the

Cabinet, would have had major implications for the viability of the IIMs, which had contributed significantly to improving corporate governance and competitiveness. The decision of the ministry led to widespread protests by the IIMs and other educational institutions. The ministry then announced that, in order to ensure the financial viability of the IIMs, the government would provide adequate direct subsidy to cover the difference between the cost of providing an education and the amount that the IIMs were going to be allowed to charge by way of fees. In other words, the government was prepared to subsidize even those students who could afford to pay higher fees in order to impose 'price controls' on the IIMs!

However, before the above decision could be implemented, there was a change in government. In May 2004, the new minister decided to reverse the earlier decision. This was widely welcomed by the IIMs as well as other educational institutions. Thus, diametrically opposite decisions with serious implications for the future of higher education in India were announced by the same ministry within a space of two months under two different ministers!

After some time, in April 2006, another major controversy erupted in the field of higher education following the announcement by the concerned minister that, in addition to reservations for Scheduled Castes (SCs) and Scheduled Tribes (STs), the government had also decided to introduce quotas for Other Backward Classes (OBCs) in institutions such as the IIMs, the Indian Institutes of Technology (IITs) and medical colleges. The decision to exercise the general powers available to the government to impose quotas with immediate effect was also announced without the formal prior approval of the Cabinet, and was publicly opposed by another Cabinet minister. The divided viewpoint within the Cabinet had the effect of making the entire admission policy highly uncertain for the students and the faculty of some of the best-known institutions in the country for quite some time until the policy was finally approved by the Cabinet.

These and similar other cases, where particular ministers have decided to announce important government decisions on their own, have now become matters of considerable public concern. The confusion has been further compounded by the disparate ideologies of the constituent parties in 2005 at the Centre as well as in the states. There have also been cases where ministers in office have publicly expressed their disengagement with important Cabinet decisions (for example, in respect of the Union government's hasty decision to dissolve the state assembly of Bihar in 2005). Despite their disagreement with a formal Cabinet decision, contrary to parliamentary norms in mature democracies, the concerned ministers did not resign as members of the Union Cabinet.

For the future governance of the country, an important issue that requires consideration is: if, on the one hand, the Cabinet cannot be assumed to be collectively responsible for ministerial pronouncements and, on the other hand, ministers cannot be held individually responsible, then who should take responsibility for the actions taken or not taken on behalf of the government? This question has become even more pertinent in the light of the diminishing role of parliament in enforcing the accountability of the Council of Ministers.

## The Politicization of Administration

As mentioned earlier, a great deal has been written on the non-accountability, corruption and inaptitude of the Indian administrative system. In addition to academics, outside observers, international agencies and the general public, a number of civil servants have also written their memoirs or recounted their experiences after their retirement from the highest offices of the state. There is now almost complete unanimity that, despite having some of the best and brightest persons in the civil services, the system as a whole has become non-functional, and that there is very little possibility of reforming it. This situation has arisen despite a great deal of trying, including efforts made at different levels by distinguished committees, commissions, associations and public-spirited persons. These efforts have come to naught because the system is dominated by internal conflicts of interest (for example, among separate trade unions for different classes of government employees), political interference, outdated statutory provisions, complicated seniority-bound procedures, fiscal stringency and the proliferation of administrative agencies that operate at cross purposes without any clear division of work.

This was not always so. Indeed, for many years after Independence, India's civil services were regarded as exemplary among developing nations. Under India's system of public administration, there was supposed to be a clear division of role between the permanent civil service and the political leadership. The bureaucracy was subordinate to the elected politicians, who were chosen by the PM at the Centre (and by the chief ministers in the states) to head different ministries and departments. The government's priorities and its work programmes were set by the elected politicians, and the bureaucracy was supposed to ensure that this programme was implemented according to the laws in force and in line with approved administrative procedures. While implementing the programmes set by the Cabinet and the ministers, bureaucrats were expected to act without fear or favour and to ensure that the benefits of the programmes flowed to the people regardless of their political affiliations. While the elected politicians were free to overrule the advice rendered by civil servants, the advisory functions of the bureaucracy were expected to be performed without regard to their impact on the private interests of politicians and the party in power.

Over the years, slowly but surely, the role of the bureaucracy has unfortunately been seriously compromised. Thus, according to the report of the National Commission set up in 2002 to review the working of the Constitution, arbitrary and questionable methods of appointments, promotions and transfers of officers by political superiors have led to the corrosion of the moral basis of its independence. It has strengthened the temptation in services to collusive practices with politicians to avoid the inconvenience of transfers and for officers to gain advantages by ingratiating themselves to political masters.

Over time, however, the politicization of the bureaucracy gathered further momentum as a result of governments pursuing their private or party interests in the guise of promoting the larger public good. Any party that comes to power is inclined to appoint favoured bureaucrats in sensitive positions, people who are expected to carry out the wishes of its party leaders, irrespective of their merits or legality. If a bureaucrat does not comply, he/she is likely to be transferred immediately to another position in another location. In one year alone, in the state of Uttar Pradesh (when there was a six-monthly rotation of the government headed by the leaders of the two parties in a coalition, the Bharatiya Janata Party and the Bharatiya Samajwadi Party [BSP]), there were 1,000 transfers among the members of the elite Indian Administrative Service (IAS) and the Indian Police Service (IPS). Under one head of government, transfers of officers, including those from the IAS and IPS, ran at an average of seven per day. Under the second head of government, who took office after the expiry of six months, transfers at different levels rose to 16 per day!

There were substantial deleterious effects of frequent transfers on the morale and effectiveness of top civil servants. The costs to the country in terms of the loss of quality of administration have also been significant. Administration has become increasingly weak and arbitrary since there is no time available to a newly appointed civil servant to acquire even the minimum knowledge necessary for an effective discharge of functions. Incompetence at the top leads to acts of passive resistance and delays by subordinates. Corruption becomes unavoidable, both to avoid transfers as well as to secure remunerative postings by corrupt officials.

As a result of the above developments, the administrative system, with multiple agencies involved at different levels in implementing government programmes, has largely become ineffective and non-responsive. The common experience of all citizens who have to deal with a government agency for any purpose, large and small, is that of insuperable problems and delays. There is also a large diversion of funds from the intended purposes to bureaucrats, politicians and middlemen at various levels of the administrative hierarchy.

A host of recommendations for improving the system has been made by numerous high-powered committees. However, the general view among experts and experienced civil servants seem to be that the reform of the system is not feasible. This is not because the country does not know what to do but because of political resistance to the reform of the civil service.

The worst sufferers of the politicization of administration are the poor because of their dependence on public services and government programmes for various facilities, such as subsidized food and health services. Unfortunately, the poor also face the maximum degree of indifference and harassment from government staff in securing access to their entitlements. Thus, a survey by the Public Affairs Centre found that, in Delhi, the average slum-dweller needed to make six trips to a government agency to resolve a problem, and that in only 6 per cent of the cases was his/her problem attended to.

The insensitivity of the administrative system to the needs of the poor, even to the urgent necessity of preventing starvation, has been confirmed by first-hand surveys and several reports by journalists and NGOs. One such survey revealed that in remote villages, children were dying of hunger despite the country's godowns bursting at the seams with public stock of food, and despite more than ₹40,000 crore being allocated by the central government for expenditure on anti-poverty schemes. Similarly, it was found that many villages were without water even after the rains because the water channels and tanks had fallen into disuse as the government had announced its intention of providing piped water out of taps. While some construction work had been started to provide tap water, the project was, however, left unfinished. As a result, the villagers had no access to either tap water or old-style tanks or wells within a reasonable distance.

The indifference of the administrative system towards the poor in providing them with their legitimate entitlements is the principal reason for the increasing disparities between urban and rural areas as well as the widening in income levels of different classes of citizens. The poor and the unemployed are more dependent on the government than is the case with other sections of the people, particularly those who are employed in organized sector and/or have access to services provided by other non-governmental sources.

In addition to control over the services provided by the government, another fertile area for reaping political benefits is the control over public-sector enterprises. Many crucial sectors of the economy are dominated by public enterprises, for example, railways, airports, public transport, oil, steel, coal, banking and insurance. For nearly four decades after Independence, many of these sectors were also characterized by widespread controls and shortages. The powers of issuing licences and allocating distribution channels for goods and services to beneficiaries (e.g., petrol pumps) were enjoyed by political authorities in charge of different ministries.

Over the last two decades, most of the controls over the economy have been removed and shortages of various kinds have also largely disappeared because of the abolition of import quotas, reduction in monopolies and the entry of new producers. Nevertheless, given the large role of public enterprises in the economy, the control of such enterprises still confers substantial powers to ministers-in-charge in dispensing political patronage to the suppliers and buyers of various kinds of goods and services. In theory, the administrative power over these enterprises is expected to be exercised by their management under the overall supervision of the boards of directors. In practice, however, the ministries exercise substantial visible and invisible control over their functioning. Large contracts for new projects also require ministerial approval after all other technical and procedural formalities have been completed. Ministries have the final say on all policy matters, for example, pricing policy or financial policy, including the issue of additional shares to the public.

## Separation of Powers

Earlier, a reference was made to the disputes that arise from time to time about the relative boundaries of the powers and jurisdictions of the three organs of the State, namely, the legislature, the executive and the judiciary. Such disputes tend to become more frequent when the political interests of the leading parties in a coalition or their leaders are under threat. While the Constitution broadly defines the jurisdiction and responsibility of each organ of the State, in disputed cases involving the legislative or the executive branch, the final-level decision is left to the judgment of the judicial branch.

In a parliamentary system of government, members of legislatures as well as of the Cabinet are directly, and in some cases indirectly, elected by the people. Members of the judiciary, on the other hand, are unelected and do not necessarily represent the 'will of the people'. The right to pass legislation belongs to the legislature, while the executive functions are supposed to be carried out under the direction of the Council of Ministers. In cases of dispute over the jurisdictional boundaries of the three branches of the State, the legislature can, with some justification, argue that it is the supreme law-making body and that the courts should not pass verdicts that have the effect of changing the legal position as approved by the representatives of the people. In defence of this position, it can be further pointed out that the courts can also be wrong. Thus, some court rulings in the past were wrong in law and had to be overturned by subsequent rulings by a larger bench of the Supreme Court. No court, not even the Supreme Court, therefore, can be considered to be infallible.

It is true that, *prima facie*, some of the past Constitutional judgments were indeed protective of private interests. For example, soon after Independence, in 1951, several court rulings overturned land reform measures as being violative of the fundamental rights of land owners. The government, led by Jawaharlal Nehru, had to amend the Constitution to implement land reforms, which were considered vital for the country's economic and social progress. Similarly, in 1970, the Supreme Court had also ruled against the nationalization of banks undertaken by Indira Gandhi's government. Special legislation had to be passed by parliament to make bank nationalization possible. A number of other instances could be cited where the judgment of the Supreme Court and other courts was not in line with popular expectations.

While all these arguments have some validity, keeping in view the recent political developments at the Centre and in the states, on balance, the long-term interests of the public and the ordinary citizen are safer when the Supreme Court continues to be the watchdog of India's democratic conventions and the final arbiter of the Constitutional validity of any law or action approved by the legislature or the government of the day. This view is not meant to detract from the merits of the parliamentary system of government in unifying India and giving people the freedom that they cherish.

At the same time, it has to be recognized that it is not prudent or politic for the ordinary citizen of India to confer supremacy on legislatures without accountability. The legality of executive action must continue to be subject to judicial scrutiny, however high the level at which such decisions were taken. Ministries have become increasingly subject to the unilateral policy preferences of individual ministers who happen to be in office at a particular point of time. There is seldom any worthwhile public debate or constructive dialogue on matters of long-term public importance. The issue here is not whether the decision taken by a particular minister on behalf of his ministry is right or wrong. The issue here is one of long-term public interest. If a minister is able to turn the economic or social policies of the country one way or the other without adequate discussion and accountability, what stops him from passing laws or rules that have an adverse long-term impact on the welfare of the people as a whole, or sections of the people who are not aligned to his party?

As mentioned earlier, the Supreme Court and other courts are not infallible, and have also given judgments that had to be reversed in subsequent hearings. From the point of view of the average citizen, the great advantage of the judicial review of decisions taken by the executive and the legislature is that everyone, irrespective of his/her beliefs and political affiliations, has access to the courts. This is not the case in respect of the executive or parliament or state legislatures. The free media does play a constructive role in enforcing a degree of accountability on the part of the government, but that by itself is not enough.

The other advantage of the judiciary being the arbiter of the legality or otherwise of an executive or legislative decision is that, even if a particular verdict is wrong or socially unacceptable, it is subject to review and reversal. This is usually not the case with legislative or executive decisions unless the government so decides. A citizen has no legal right to ask for a review of the decisions taken by the legislature or the executive, even if these are not in the public interest. The Right to Information Act, approved in 2005, is an important step forward in making the executive accountable to the people directly. However, in case of any unjust or partisan decisions taken by the government, the remedy would still lie with the judiciary.

Keeping in view India's experience as well as that of other democracies, it is clear that a rigid demarcation of legal powers among the different branches, irrespective of the specific circumstances, is neither feasible nor desirable. By and large, under normal circumstances, it is appropriate for the different branches of the state to work in harmony and confine themselves to their primary tasks as enshrined in the Constitution. Parliament as the highest representative body should have the unquestioned authority to pass laws that it considers appropriate. The executive branch should be accountable to parliament, and should have full administrative powers to implement laws and programmes as approved by parliament. And the judiciary should give verdicts and settle legal disputes as per the law of the land. Most of the time, in all mature democracies, there should also be no cause for jurisdictional conflict among different organs of the state.

However, there are times when sectional interests and the 'compulsions of coalition politics' can become the primary drivers of the laws passed by the parliament and/or the administrative actions taken by the executive. Some of these laws and executive decisions may run counter to the intent of the Constitution and adversely affect the fundamental rights of the people. If the political majority in parliament is fragmented, and there is a serious conflict, ideological or otherwise, among coalition partners, some parliamentary decisions may reflect sectional electoral interests rather than the long-term interests of the people as a whole. There may also be times when the political or private interests of a leader of a ruling party or a group of leaders may be under public scrutiny because of certain exogenous or endogenous developments. In such exceptional, and hopefully infrequent, circumstances, it is necessary to have a court of last resort for deciding on the Constitutional validity of specific laws or actions initiated by the legislature or the executive. The final legal arbiter in such cases can only be the judiciary, which is directly accessible to the public, and whose verdicts are in any case subject to review and appeal.

It is, of course, true that the judiciary can also make mistakes. In view of the enormous delays and multiple levels of appeals, it can also be argued that the judicial system itself is in urgent need of reform in order to provide speedy justice. However, even after taking all these imperfections into account, there is no doubt that, on balance, the country is better off with the judiciary as an additional checkpoint on the legality of actions taken (or, for that matter, not taken) by the legislature and the executive. It should be free to issue appropriate directions to any agency of the state if its actions are considered arbitrary, partisan and violative of the intent of the Constitution in order to give India a government of the people, by the people and for the people.

# 10

## THE CRISIS OF GOVERNANCE

### 2005

The crisis of governance in India, and the apathy of the governance structure towards the welfare of the general public, needs no introduction. If we look back at the period around 2005, India had been in the news for its economic performance. It had also emerged as a leading exporter of software services and other high-technology exports. At the same time, despite its high growth potential, India also had the world's highest number of persons below the poverty line. The estimates of the number of poor persons in India varied depending on the methodology used for estimation. However, even the most conservative estimates released by government agencies showed that as many as 300 million Indians were below the poverty line, and did not earn enough to ensure even the minimum intake of food and nutrition. The conditions prevailing in its urban slums and rural areas were among the worst in the world, and even the most optimistic observers did not foresee any possibility of a dramatic improvement in the near future.

Over the years, the government had launched a large number of programmes which were 'targeted' to remove poverty through the creation of jobs, the provision of subsidized credit to the poor or the delivery of free food in exchange for work. These programmes had no doubt benefitted the poor, and helped to reduce the extent of poverty; however, all field studies as well as casual observation suggested that leakages in government-funded anti-poverty programmes were very high. In the late 1980s, leakages in these programmes were as high as 85 per cent, according to the then PM, Rajiv Gandhi. Since then, the position was likely to have become worse rather than better because of political corruption and administrative ineptitude.

Ironically, a substantial part of the funds allocated for anti-poverty programmes in the annual budget also remained unutilized or was diverted by state governments and local authorities to meet other revenue expenditure. According to an expert commission set up by the Supreme Court of India (in response to a Public Interest Litigation), even Maharashtra, which had a well-established administration, failed to utilize as much as 78 per cent of the allocation of funds made by the Centre for providing nutritious food to children as part of the PM's Gramin Yojana scheme. Instead, it demanded funds from a large number of parents of poor children to fulfil the state's obligations. The position was worse in several other states, such as Bihar, Jharkhand and Uttaranchal, which were reviewed by the expert commission. In Jharkhand, for example, the government had failed to avail of the entire budget allocation because financial sanction could not be issued on time.

It was equally shocking that out of the relatively small budget expenditure on anti-poverty programmes, as much as 70–80 per cent was on account of the salaries of government servants at various stages of implementation of these programmes. The same was true of subsidized credit provided by the refinance agencies and banks owned by the central government. Thus, subsidized credit provided at 6 per cent per annum by the central refinancing agency had to pass through state, district and primary cooperative credit agencies before it reached the farmer. The intermediation costs were more than double the initial interest rate of 6 per cent charged by the central agency, and the cost of credit to the farmer at the delivery point was 14 per cent or higher.

## The Poor in 'Shining India'

The dichotomy between a 'shining', fast-growing India and its persistent poverty is certainly a puzzle that has baffled many development economists as well as ideologues. The answer to this puzzle does not lie in the proposition, advanced by some economists and central planners of the old school, that there is an inherent conflict between the objective of raising the growth rate of the economy and that of reducing poverty. As it happens, the global experience is that countries and regions that have registered high and sustained growth rates over a reasonable period of time are also the ones that have achieved the best results in reducing poverty and improving the health and nutrition of their people. In some cases, the progress in reducing poverty or improving the level of human development indicators has no doubt been much greater than would seem warranted by their growth rates, as has happened in Kerala and Sri Lanka.

There have also been cases where high growth were combined with worsening of the poverty ratio, as in Brazil in the 1970s, or where high per capita incomes did not result in adequate progress in education and other social services, as is the case in some oil-rich countries. However, such cases are not many and they have their own special reasons. Over time, it also became evident that Sri Lanka and Kerala, which despite low growth had made commendable progress in poverty alleviation, later found it difficult to sustain the process. Per capita expenditures on anti-poverty programmes declined because of fiscal stringency. And with low industrial growth, unemployment became a pervasive problem. This was a major obstacle to further progress on the poverty front.

It is obvious that poverty alleviation in a low-income country with poor basic amenities and poor availability of essential public services (such as primary education, water, power and transport) is feasible only if the government has the financial capacity to create the necessary infrastructure for the provision of such services to the poor. It is also likely that the higher the growth rate of the economy, the higher is the growth of government revenues and its capacity to finance social expenditure. Whether the government actually does so or not is naturally a matter of public policy. However, a low growth rate is not pro-poor. Nor does it help the debate on social or public policies. It is legitimate to ask for more government expenditure and more government intervention in favour of the poor or for more pro-employment growth policies. But it is fallacious to argue that the government can be more pro-poor in a stagnant or low-growth economy for any length of time.

The real answer to the puzzle about India's high growth combined with persistent poverty in the early 2000s—and also later—lies in what can perhaps be described as the growing 'public–private' dichotomy in economic life. It is a striking fact that economic renewal and positive growth impulses have been occurring largely outside the public sector—at the level of private corporations (for example, software companies), autonomous institutions (for example, IIMs or IITs) or individuals at the top of their professions in India as well as oversees. In the government or public sector, on the other hand, there has been a marked deterioration at all levels—not only in terms of output, profits and public savings, but also in the provision of vital public services. Fiscal deterioration and the inability to provide essential services are, without doubt, intricately linked. In India, most of the public resources are exhausted in paying salaries or interest on debt, thus leaving very little resources for the expansion of public/publicly supported utilities in vital sectors.

The widespread and persistent poverty, despite high growth in the private sector and some parts of the public sector (such as the oil companies, where the government has a relatively strong monopolistic position) can only be explained by the inability of the administrative structure, consisting of ministers and civil servants, to manage resources efficiently and deliver public services without massive leakages. Let us briefly recap the problems that bedevil the public delivery systems:

- India can be proud of the fact that the rule of law prevails in the country, and that even the mightiest are not above the law. Although the delays in the judicial process are nothing short of legendary, there is still widespread respect for the rule of law. However, for historical reasons, it is also a fact that the legal system provides full protection to the private interests of the so-called 'public servants' often at the expense of the public that they are supposed to serve. In addition to complete job security, any group of public servants in any public-sector organization can go on strike seeking a raise, promotions and bonuses, irrespective of the costs and inconvenience to the public. There is very little accountability of the public servant for non-performance of duty. This is despite a Supreme Court judgement in 2003 against strike by government servants as no penal provisions have yet been prescribed by the government.

- The authority of the central and state governments to enforce their decisions has worn out over time. Governments can pass orders, for example, for the relocation of unauthorized industrial units or other structures, but implementation can be delayed if they run counter to the private interests of some (at the expense of the general interest). Similarly, governments may decide to restructure public utilities to curtail waste or output losses, but these decisions do not necessarily have to be implemented if they adversely affect the interests of public servants employed in these organizations.

- Governments at different levels may announce plans and programmes to provide social services (such as expanding literacy), but these initiatives are unlikely to be implemented on the grounds of fiscal stringency. For example, in 1994, the Tenth Finance Commission projected a rate of growth in real terms of 2.5 per cent for expenditure on elementary education up to the end of the century for four states where the incidence of poverty and illiteracy was among the highest in the country. This projected rate of growth in expenditure was lower than the growth of the population in the relevant age group, and grossly insufficient to cover new programmes for adult illiterates. Ten years from the Commission's report, it was interesting to note that the real expenditure on elementary education, outside of salaries of government teachers, in these four states had actually been negative!

- The processes and procedures for conducting business in government and public-sector organizations, over time, have become non-functional. There are multiple departments involved in the simplest of decisions, and administrative rules generally concentrate on the process rather than the results. There is very little decentralization of decision-making powers, particularly financial powers. Thus, while local authorities have been given significant authority in some states to implement national programmes, their financial authority is limited. Transfers to local authorities for health spending, for example, average less than 15 per cent of state government budgets.

- The multiplicity of functions and responsibilities placed upon ill-equipped and ill-trained staff in public offices and local institutions makes it almost impossible to deliver services with any degree of efficiency, particularly in rural areas. For example, a 'multipurpose' female health worker is required to perform as many as forty-seven tasks, to be undertaken on a regular basis!

To improve governance and provide better services to the public within a democratic framework, it is necessary to impose greater accountability on both ministers and civil servants. Some of the essential elements of reform are discussed below.

## The Myth of Collective Responsibility

Each department and ministry of the government is headed by a politically appointed minister, who is an MP and represents the party in power (which has a majority in parliament either on its own or as a part of a coalition government). The minister is the chief executive of the ministry, and reports to the PM. In a parliamentary form of government, the Cabinet is supposed to have a collective responsibility for all the decisions taken by each ministry either on its own or with the approval of the Cabinet. Each minister, as a member of the Cabinet, is also answerable to parliament. The MPs have the right to ask questions, move call attention motions, introduce resolutions and demand accountability for all the decisions taken by ministers and the performance of their respective ministries.

If we look back at 2005, India had one of the lowest per capita income levels in the world and the highest number of poor persons. Therefore, alleviation of poverty, creation of employment opportunities and provision of better public services to the poor were among the principal items on the economic agenda of every political party at the time of elections. The instrumentalities and specific policies proposed to be adopted, if voted to power, may have varied from one party to another, but the anti-poverty objective was the same. In view of this, it was surprising that with the different combinations of parties that had formed governments in the preceding 15 years, the public delivery system had continued to deteriorate. While there may have been numerous questions, calling attention motions and resolutions in parliament on unemployment and the non-availability of basic infrastructure in rural areas, no minister had actually been held accountable (or censured) for the poor performance of his/her ministry in these vital areas.

The reason for parliamentary inertia and the non-accountability of ministers for the performance of their ministries lay in the supremacy of the parties and their leaders in the political system. As long as the party or the combination of parties had a majority in parliament, ministers were supreme because they enjoyed the patronage of the leader of the party or the PM. No resolution or calling attention motion could be adopted in parliament without majority support, and no harm could come to the minister or his/her special interests or political constituency if his/her party was in majority. The position becomes worse in a coalition government with a thin majority and the disparate ideologies of constituent parties. In this case, ministers belonging to a party other than that of the PM are not even accountable to him or to the Cabinet formed by him. In the 15 years since the 1990s to the first decade of the 2000s, there have been a number of governments (including the government that came to office after the general elections in 2004) where the Cabinet had been headed by a party with less than one-third of the total number of seats in the Lok Sabha. The survival of the government thus depended crucially on the continued support of other parties, large and small, local and national. It was not surprising, therefore, that the concept of ministerial responsibility for the performance of ministries or the government as a whole had become largely illusory.

The fractured electoral verdict in the 2004 Lok Sabha polls—with no national party securing even one-third of the seats and the emergence of a large number of local parties as pre-election allies—only reinforced the scenario. The government was expected to continue having some excellent ministers with a deep commitment to the public good. However, on the whole, the sense of collective responsibility and the accountability for the management of public resources and delivery of services to the poor was likely to be absent. This ground reality could not be denied.

Assuming that political parties, the civil society and the enlightened members of the Indian public were serious about removing the worst forms of poverty and deprivation, a new institutional initiative was urgently required to enforce ministerial responsibility for the efficient delivery of public services and anti-poverty programmes all across the country. This could be achieved only if the cherished doctrine of 'collective' responsibility for all actions of the government was replaced by the notion of 'individual' responsibility of ministers for implementing programmes that were of direct concern to the poor. The doctrine of collective responsibility could continue to prevail for all other political purposes, including the continuation of a government in office.

The objective of better delivery of public services can be achieved if quantitative annual targets are agreed upon with each of the concerned ministries at the time of the annual budget and each minister is made responsible to parliament for achieving those targets. While quantitative annual targets were already being worked out as part of the annual plans of ministries, they had no sanctity or force of ministerial responsibility behind them. These targets were primarily for obtaining higher budgetary allocations. However, there was no accountability for the actual implementation or achievement of agreed targets. One way to counter this issue is that the government be made responsible for placing before parliament a report on actual achievements in relation to the agreed targets. This report should be the focal point of discussion in parliament on a ministry's budget, and if there is a shortfall of more than the agreed percentage (say, 15 or 20 per cent), then the minister must be held responsible, and expected to relinquish ministerial office for at least one year. If there is a change of ministers during the course of the year, then the new minister must once again affirm or change the target with the approval of parliament.

One obstacle in assigning individual ministerial responsibility for realizing public service targets is that a number of other ministries, in addition to the administrative ministry, are involved in the actual implementation and approval of budgetary expenditure. The Ministry of Finance may be involved in the approval of the actual expenditure, even if the necessary expenditure is already included in the budget. Yet another ministry may be involved in the design of a particular programme if that programme also cuts across its area of administrative responsibility. It may, therefore, be considered unreasonable to hold a particular minister responsible for failure to realize the announced public service targets, as the reasons for non-performance may lie elsewhere. Under the present system, where there is substantial diffusion and overlapping of administrative responsibility, there is certainly merit in this argument. However, the answer to this problem lies in reforming the administrative system rather than in denying individual ministerial responsibility for implementation of an approved programme. Henceforth, it may be stipulated that all concerned ministries, including the ministries of finance and planning, would be appropriately consulted at the time of fixing the annual target rather than at the stage of implementation. If budgetary resources are not available or the design of the programme is not yet approved, the announcement of the anti-poverty and other targets should also be deferred. Once the target is announced by a ministry, it should have the full authority to implement it and it would be the only ministry which is held accountable for actual performance.

The proposal is certainly unconventional and contrary to established parliamentary practice, which does not recognize individual ministerial responsibility for a ministry's performance. However, this is the only feasible option if the country is serious about reversing the deterioration of public services due to financial leakages and ministerial apathy. It would be desirable if, with the full support of all the parties in the government and the Opposition, a binding resolution to this effect is adopted by parliament as early as feasible. A higher degree of direct ministerial responsibility for performance, along with a restructuring of the administrative system is now essential if India is to succeed in abolishing the worst form of poverty, by say, 2025.

## The Failure of Administration

The actual delivery of public services is obviously dependent on government officers and employees who occupy administrative positions at different layers of the hierarchy in the delivery process. There are senior officials in various ministries at the Centre who are responsible for deciding the policy, the amount of budgetary grant or loan to be made available to state governments, and the implementing agency responsible for carrying out the project. At the state or agency level, there are officials in the main secretariat or elsewhere who have to disburse the funds and take various other administrative decisions. Then there are district-level officials who are responsible for monitoring the progress of the programme, making site visits and filing returns. Finally, there are ground staff at the primary level who are actually responsible for delivering a particular service to the public, removing deficiencies, attending to complaints and levying a token fee, if necessary. If there is a problem or deficiency at the field level (say, the non-availability of medicines because of lack of funds), it is likely to be referred to the district, state or central level, or all three, for resolution. In most cases, particularly those involving staff or funds, officials in multiple agencies at all three levels are likely to be involved.

India's civil services are overburdened by an imbalance in the skill mix. In 2005, nearly 93 per cent of the civil service comprised of the so-called Class III and Class IV employees in the Government of India. The percentage of such employees in state governments is even higher. They include frontline delivery workers, but the overwhelming numbers are clerks, typists, messengers, peons and sweepers. The over-abundance of support personnel often exists alongside chronic shortages of skilled staff in essential services, such as water supply, health, sanitation and schools. To make matters worse, the civil service is divided into dozens of cadres, each with its own terms and conditions of service, with controlling authorities wisely dispersed among various departments. Some staff, particularly senior staff, are transferred after short tenures (ranging from a few days to less than a year), but other staff cannot be transferred between departments or locations irrespective of functional requirements.

The administrative cobweb, with multiple agencies and no clear-cut demarcation of functions, creates insuperable problems and delays for all those who have to deal with a government agency for any purpose, large or small. At least six or more central ministries are involved in controlling or regulating every sector of the economy. Thus, matters concerning the food processing industry normally have to go through nine ministries—agriculture, food and consumer affairs, health, commerce, food processing, rural development and others. In 2005, the Ministry of Human Resource Development (HRD) at the Centre (which was earlier called the Ministry of Education) consists of four separate departments: the Department of Education, the Department of Culture, the Department of Women and Child Development and the Department of Youth Affairs and Sports. Each of these departments has several sub-departments or divisions. The Department of Education has separate units concerned with primary education, secondary education, technical education, teachers' education, higher education, book promotion and copyright, planning, languages, district education and international relations, among others.

Similarly, the Department of Women and Child Development and the separate Department of Youth Affairs and Sports have several divisions concerned with different aspects. Departments have several commissions and subsidiary organization under them reporting to different divisions and authorities with cross-cutting functions. While there is a multiplicity of specialized departments, divisions and organization, none of them has adequate authority to take any decision concerning even the most non-controversial items. For example, the matter of providing better sports facilities for young women in schools would require consideration by practically all the divisions and organizations as it concerns women, sports, education, youth and, perhaps, welfare! After the matter has gone around the ministry's various units (with conflicting views and objectives) over several months or years, it would need to be considered by an inter-departmental committee of the ministry, then a number of secretaries, and finally the minister, all of whom are likely to have the most casual acquaintance with the subject. Thereafter, of course, if finance, law, planning, security and any other aspects are also involved, the matter would need to be referred to other ministries with their own multiple layers of departments and divisions.

The burden of weak administration naturally falls mainly on the poor because of the indifference of government staff to them. In Delhi, the average slum-dweller needed to make six trips to a government agency to resolve a problem, and his/her problem was attended to in only 6 per cent of the cases. Other households had to make four trips, and in their case, the rate of success was substantially higher, but still only 27 per cent.

The insensitivity of the administrative system to the need of the poor, even to prevent starvation, has been confirmed by first-hand surveys and reports by journalists and NGOs. In remote villages, children were dying of hunger despite adequate availability of public stocks of food. On further investigation, it was found that the vast network of officials set up to take care of the interests of the poor was in denial. Similarly, it was found that many villages were without water even after rains because water channels and tanks had fallen into disuse as the government had announced its intention to provide piped water out of taps. While some construction work had been started to provide tap water, the project was left unfinished. The villagers had access to neither tap water, nor old-style tanks or wells within a reasonable distance.

This is a long litany of woes. It is important to understand the complexity of the existing system, its historical evolution and its past failures to work out a possible solution. So far, the effort has largely been confined to reforming the system from within by issuing circulars for expeditious delivery, decentralization of some functions to panchayats (without delegation of financial powers), creating new specialized departments (with existing staff skills) and greater monitoring. In addition, some effort has also been made to make the system more accountable by providing the public with the right to seek information and easier access to credit for self-help groups and micro-credit organizations. As a result of these initiatives, in some states, there has been some improvement in the delivery system. There are also some organizations of the government (for example, the Department of Space, the Telecom Department, the National Highway Authority and Delhi Metro) which despite all the staff handicaps have achieved excellent results.

However, these are exceptions—welcome as they are—to the general rule of administrative decay in the functioning of the government. As in the past, there are individual persons at high political or administrative levels with integrity and commitment to the public good. However, given the dominance of special interests in the political parties as well as the civil services, it is unlikely that the administrative system can be improved from within.

While efforts to make the system more accountable must continue, an important policy measure that the government can take is to reduce its role in the governance system at different levels. Take, for example, the reform of the licencing system in telecom. Earlier, these licences were service-specific, user-specific, technology-specific, area-specific and vintage-specific. This wide differentiation in the types of licences required constant state intervention to remove the difficulties faced by operators and consumers. It also led to inefficient services and widespread corrupt practices by suppliers as well as supervisors. This unwieldy system was replaced, after a great deal of effort, by a universal access licence which was made freely available at a pre-announced fee. This reform reduced the role and functions of the State, and also improved services to the consumers.

A similar system deserves to be followed in all sectors of the economy. In addition, it is also necessary to make a distinction between policy direction (that is, laying down policy guidelines and monitoring performance) and the actual implementation of programmes. In addition to lack of adequate resources, the most important problem in the public delivery system is the mismanagement and theft of the available facilities and resources. For example, in respect of as essential a service as drinking water, if distribution and transmission losses were reduced by even half through better management of the available capacity, the improvement in the supply of services and financial savings would be immense.

International experience in the management of public services shows that the objective can be achieved if a distinction is made between the ownership of these services (by the government) and the delivery of such services (by private and local enterprises). Case studies from several countries have shown that in every case where the management of a public service was contracted out to private enterprises, the distribution and quality of the service improved and the net cost to the public was reduced. What is more, a large number of jobs and many new enterprises were also created. In most cases, the public authorities retained the responsibility for regulating and monitoring the activities, providing subsidies where necessary and laying down distribution guidelines. In India, two noteworthy examples of public–private collaboration in the area of public services are the public call offices (PCOs), which revolutionalized the availability of telephone services all over the country in the 1990s, and the Sulabh Sauchalayas, which are estimated to have provided sanitation facilities to ten million people at very low cost.

In several countries, the model of public–private partnership (or 'micro-privatization' under public supervision) has replaced the old system of public ownership and public delivery in certain important sectors. In India, in respect of telephones as well as sanitation services, the new initiatives were supplemental to the public-sector facilities. In other words, they did not replace the public-sector organizations responsible for the delivery of these services. In view of entrenched political and bureaucratic interests as well as for practical reasons (to avoid the disruption of existing public-sector services), this was a wise decision. The supplemental approach expanded the availability of services and created a more competitive environment without affecting government employees and raising resistance. It is now necessary to adopt a similar approach in respect of all essential public services. No new facilities or employees should be added in the public sector, and additional budgetary allocations (over and above the existing salary and maintenance expenditures) at the Centre, states and local levels should be made for the delivery of services by private enterprises, including non-government professional organizations.

At the same time, it is of utmost importance that a civil society movement is launched for reform of the civil service without delay. This issue needs as much attention as is given to several other vital issues of public concern, such as freedom of speech, gender equality, reservation of seats for representation of women in legislatures and so on. There is political resistance by governments for reform in all these areas. However, with popular support, India has been able to make at least some progress even on the most contentious issues. The question of civil service reform and better governance should now figure high on the public agenda.

The most critical issue that needs to be tackled is the 'motivational' or morale issue at the higher levels of civil services. Civil servants, who join government service through a competitive process, are generally highly competent and motivated when they enter the service. However, after a few years of service, there is a perceptible decline in morale, commitment and efficiency. An important reason for this decline is the power available to politicians to harass a civil servant who does not abide by their wishes. The easiest method of doing so, which is widely used by ministers, is to frequently transfer civil servants at very short notice. After a few such transfers, cynicism sets in as most civil servants would rather abide by ministerial wishes than put themselves and their families through the inconvenience caused by yet another transfer. Civil servants at higher levels, who technically pass the orders of transfers, are also reluctant to intervene. A pernicious development is that transfer decisions of even lower-level officials, which used to be taken earlier by the civil service itself, are now taken by ministers. The common belief in the civil service is that in order to survive, ministers and their officers (however corrupt or incompetent) must be kept in good humour.

In order to improve the morale of the civil service, the ministerial powers to transfer civil servants without adequate cause need to be moderated as early as possible. Except at the highest levels, say the first two rungs of the service (secretary and additional secretary), who have to deal directly with ministers, the powers of posting and transfers should be entirely within the jurisdiction of the civil service. An appropriate mechanism (such as the public services boards) is already in place, and they should be given full and final powers for posting of civil servants up to the designated level. For the highest levels in the service, postings and transfers may continue to be subject to the approvals of the political authority. However, once a civil servant has been appointed to a particular post with due approval of the government, his/her further transfer by the same government before a stipulated period (of, say, three years) should be effected only for non-performance or lack of integrity. The reasons should be stated in writing, with appropriate documentation justifying the decision, rather than left vague. The officer should have access to the reasons for transfer and should have the right to appeal to a designated higher authority.

A further measure for greater empowerment of civil service is to reform the procedure for launching vigilance inquiries and the number of agencies involved in such investigations. The ease with which investigations can be launched without adequate cause, and then closed after several years for lack of evidence, is a major cause of harassment and pain for the honest civil servant. Increasingly, there is a tendency among civil servants at higher levels to avoid taking a decision, according to the rules in place, on a financial or a controversial matter without seeking ministerial approval. In case a decision on such a matter is taken by the civil servant (who is otherwise competent or authorized to do so), it is feared that an inquiry may be launched at the instance of a minister or a business group which is adversely affected by that decision. The fear of taking decisions is a major cause of delays and atrophy in the decision-making process.

The basic issue that needs to be tackled for improving the morale of the civil service is really that of 'separation of powers' *within* the executive—between ministers and civil servants in so far as postings, transfers, promotions and other similar administrative matters are concerned. The separation of powers among the three branches of government—the executive, the legislature and the judiciary—is already enshrined in the Constitution. Although there has been considerable encroachment of the executive powers into legislative, and even judicial areas, it can still be said that these three separate branches enjoy a certain measure of autonomy (if they wish to exercise it). Within the executive branch, however, the civil service is now completely dependent on the pleasure of the ministers in regard to even the most mundane and routine administrative matters. It is essential to revert to a rule-based system of administration, which circumscribes the powers of politicians and confers greater authority on the civil service itself for self-regulation. The greater empowerment of the civil service must, of course, go hand in hand with greater accountability of civil servants for their performance and ethical conduct.

# 11

## THE ECONOMICS OF
## NON-PERFORMANCE

### 2005

At first glance, the title of this essay may seem a bit odd. In the first half of the decade of 2000–10, India was being commended for its excellent economic performance by economists, expert commentators and international agencies. It was one of the fastest-growing economies, and there was an emerging consensus that if India followed the right policies, by 2025, it would be the third-largest economy in the world. The optimism about India's growth potential was further reinforced by India's success in avoiding 'contagion' after the 1997 East Asian financial crisis, which adversely affected a number of other developing countries. In 2004, India's BOP position was stronger than at any time in its post-Independence history, and it had one of the highest levels of foreign exchange reserves in the world—amounting to $130 billion.

All this was certainly true. However, looking ahead, it was not clear whether India's current position would continue over the long run. Thus, for example, as far back as 1956, the Second Five-Year Plan was launched with great fanfare after considerable debate among leading economists in India and abroad. The Plan was supposed to bring about a transformation in the economy, make India self-reliant and abolish poverty in the following 25 years, i.e., by 1981. However, very soon, the country was engulfed in a major foreign exchange crisis and remained trapped for the next 20 years in a vicious circle of low economic growth and high poverty. Similarly, in the early 1980s, with foreign exchange reserves beginning to build up, the savings rate crossing the 20 per cent mark for the first time and the economy running a food surplus, many felt that the time for India's economic take-off had finally come.

However, before the end of the decade, the economy ran out of steam again and in 1990–91, it was caught up in yet another serious BOP crisis. The new government, which came to power in 1991, launched an impressive programme of economic reforms, which thankfully yielded results. The industrial licencing system was abolished, foreign exchange controls were liberalized, import tariffs were lowered and the burgeoning fiscal deficit was reduced. During the four-year period of 1993–94 to 1996–97, the growth rate exceeded 7 per cent per annum, and once again, there was considerable optimism about India's economic future. However, soon the economy slowed down again and in the first three years of the new millennium (2001–02 to 2002–03), the average growth rate was less than 5 per cent, partly because of the severe drought in 2002–03. The picture changed dramatically once again in 2003–04, and the growth rate was expected to exceed 8 per cent, which was the second-highest growth rate in the world after China, whose growth rate was close to 10 per cent.

It was natural to ask at that point in time—has anything really changed about India's economic prospects, or is the economy likely to continue to swing from a positive to a negative outlook? Naturally, there was no unanimous or unequivocal answer to this question. There was indeed a fundamental change in India's global economic position and opportunities for it to accelerate its growth rate were truly immense. In view of the changing role of knowledge-based services (such as professional and IT services) in the overall growth, the sources of comparative advantage of a nation were vastly different then from what they were 50 or even 20 years ago.

Very few developing countries were as well placed as India to take advantage of the phenomenal changes that had occurred in the fields of production technologies, international trade, capital movement and the deployment of skilled manpower. An important change in production technology from India's point of view was the importance of IT and software in the value of output and productivity in all sectors of the economy, including manufacturing. India had the knowledge and the skills to produce and process a wide variety of industrial and consumer products and services. Another important factor in India's favour was international capital mobility and the integration of global financial markets. Domestic savings continued to be important for the country's development. However, scarcity of domestic capital was no longer a binding constraint. Increased mobility of capital had ensured that global resources flow to countries which showed high growth and high returns. It was now possible for India to participate in the virtuous circle of higher growth, higher external capital flows, and higher domestic incomes and savings, which in turn could lead to further growth.

While there was no doubt about India's immense 'potential,' realizing the advantage of new opportunities, would require a change in the country's vision of the future and its economic strategy. Some changes had no doubt been made in this direction, particularly in the 1990s, but much still remained to be done. On present reckoning, it was not yet clear that economic performance in the foreseeable future would significantly exceed the post-1980's trend rate of growth. The performance of economy in the immediately preceding years had certainly been much better than what was achieved in the previous three decades (less than 4 per cent per annum), but it was certainly well below India's potential. If India was prepared to grasp the opportunities that were available, the trend rate of growth could be 8 per cent per annum or more. There were, however, three important factors which could impede or delay the realization of its full economic potential. These were:

I. the deadweight of the past in economic vision and strategy;
II. fiscal disempowerment, largely due to the power of 'distributional coalitions'; and
III. the growing 'disjuncture' between economics and politics in public life.

## The Deadweight of the Past

The reasons for India opting for a highly controlled state-dominated development strategy after Independence are well known. The economic profile of the country at that time was distressing. There was hardly any growth in the previous half-century, and both agriculture and industry were characterized by severe structural distortions. Like other underdeveloped countries, India was an exporter of cheap primary products and an importer of industrial products with secular decline in its terms of trade and stagnation in per capita incomes. During the first half of previous century, the growth rate of national income was at less than 1 per cent per annum, which was comparable to the rate of population growth during this period. In real terms, therefore, at the time of Independence, the average Indian was as badly off as he/she had been at the turn of the century. Against this background, there was unanimity among nationalist intellectuals, political leaders and industrialists about the preferred directions of economic strategy after Independence. The need for the government to occupy the commanding heights and to lead from the top received further support from the astounding success of the erstwhile Soviet Union in emerging as a rival centre to the West within a very short period. At that time, India played a pioneering role in giving expression to the aspirations of the newly independent Third World countries in the economic field. Following the example of the Soviet Union, there was also a broad consensus on many of the strategic issues, such as the vital role of the public sector, the discouragement of foreign investment, the development of heavy industries and the need for centralized allocation of resources.

While the reasons for adopting a centrally directed strategy of development were understandable against the background of colonial rule, it soon became clear that the actual results of this strategy were far below expectations. Instead of showing high growth, high public savings and a high degree of self-reliance, India was actually showing one of the lowest growth rates in the developing world, with rising public deficits and periodic BOP crises. In 30 out of the 40 years between 1950 and 1990, India had BOP problems of varying intensity. Looking back, it is hard to believe that for as long as four decades after 1950, India's growth rate averaged less than 4 per cent per annum, and the per capita income growth was less than 2 per cent per annum. This was at a time when the developing world, including Sub-Saharan Africa and other least developed countries, showed a growth rate of 5.2 per cent per annum.

However, the most striking failure was not in terms of growth, or even in the precarious situation of the BOP. Although the argument is not convincing, it could still be claimed that the low growth outcome was on account of a number of factors which were beyond our control, such as the border wars, severe droughts, periodic oil shocks and, finally, the inhospitable global environment. The BOP difficulties could also—with some imagination—be attributed to the global woes of primary producers, and the struggle of a poor developing country like India to industrialize and become self-reliant in heavy industry (which previously had been the monopoly of the rich industrialized countries). The most conspicuous failure for which there is no alibi, and the responsibility for which lies squarely and indisputably with the policymakers is the erosion in public savings and the inability of the public sector to generate resources for investment or the provision of public services.

It may be recalled that an important assumption in the choice of post-Independence development strategy was the generation of public savings, which could be used for higher and higher levels of investment. However, this did not happen, and the public sector, instead of being a generator of savings for the community's good, became a consumer of the community's savings. This reversal in roles had become evident by the early 1970s, and the process reached its culmination by the early 1980s. By then, the government had begun to borrow not only to meet its own revenue expenditure, but also to finance public-sector deficits and investments.

Thus, the public sector, which had a commanding presence in almost all industrial sectors of the economy, particularly heavy industry, gradually became a net drain on the society as a whole. It is interesting to note that the central government's total internal public debt reached ₹5,00,0000 crore by the mid-1990s, and nearly one-third of it was accounted for by assets held in the public sector. Interest payments on public debt at that time amounted to nearly ₹40,000 crore, which were financed by new net borrowings and represented nearly 70 per cent of the Centre's fiscal deficit.

Looking back at the performance of the public sector in contributing to national savings (which, by 2005, had been negative for the past three decades), it was amazing how much of the economic and political debate on future strategy was still conditioned by the pre-1947 colonial experience and special interests. Irrespective of which party or coalition of parties was in power, political leaders (with very few exceptions) expressed their confidence in the ability of the public sector to generate savings.

Disinvestment targets, particularly for loss-making units, were announced from time to time, but were unlikely to be reached.

The issue here was not the public sector vs the private sector or the ideological predilections in favour of state-dominated development strategy vis-à-vis market-dominated development strategy. Nor was it about the virtues of globalization or its discontents. The issue was simply about the proper uses of national savings in an environment of rising fiscal deficits. Is it appropriate to use these savings for financing further losses of public-sector units which are of no particular interest or service to the vast majority of India's poor? Is it justified to continue with vast and growing government borrowings and the disproportionate burden of interest payments on the government's budget when earlier borrowings invested in the public sector were not yielding adequate returns? There is no doubt that the financial interests of workers in the public sector, whether these units are yielding returns or not, deserve to be protected. The crucial issue is whether the most economically efficient way of protecting these interests is through further government borrowings for financing mounting losses and low returns in these units? Or, whether these interests can be adequately protected through a more productive use of the capital (including land) that is locked up in these units?

It could be argued that what was needed was better and more professional management by reducing political/bureaucratic control of public enterprises. Several government committees over the preceding three decades had made recommendations along the same lines, which had been accepted by the government 'in principle'. Some ministries had also tried to implement these recommendations, and achieved some success, albeit briefly. However, with the change in government or ministers (who generally have even shorter tenures than the government because of Cabinet reshuffles), uncertainty and drift was likely to continue. It was better to recognize this political reality rather than evade it in considering the options. Another factor which needed to be taken into account was that the high cost of production and low value added in several public-sector units was not necessarily because of management deficiencies but rather because of outdated technology, inappropriate location, non-marketable product mix and other extraneous factors. The government may have still considered it appropriate to continue with investments in some public-sector units for strategic or equity reasons (say, for development or in certain regions). However, it would have been advisable to select only those units for preservation which specifically satisfied the strategic or social objectives rather than invest further in public enterprises which were not generating adequate returns.

Similarly, while availability of capital was no longer a constraint to development in view of international capital mobility, any policy measure to liberalize FDI in India continued to attract political controversy. The suspicion with which investment was viewed was also a direct consequence of the colonial experience in the nineteenth century and the first half of the twentieth century. During that period, FDI conferred ownership and management rights on foreigners, who exploited these rights to 'drain' investible surpluses and resources out of India. For this reason, the national political movement during the pre-Independence period regarded foreign domination of Indian industry as a major cause of the country's poverty. While this was indeed the case 50–60 years prior to 2005, the situation had vastly changed. Indian industry and infrastructure were largely owned by Indians. At the beginning of the twenty-first century, the share of FDI in the total capital stock in Indian industry was among the lowest in the world and relatively insignificant in relation to the size of the economy. In 2003–04, for example, FDI was only 0.7 per cent of the national income.

A much better approach in respect of FDI would have been to specify that for the next decade (or some such period), India would follow exactly the same policy as, say, China. If, after the specified period, Indian markets were found to be flooded with foreign investment, and such investment formed a sizeable proportion of the total capital formation (which was unlikely), the policy could be reviewed. A better approach would, of course, have been to ensure that Indian markets were competitive and open, and not monopolistic with high levels of effective protection.

There were also several important policy areas where there has been a welcome and decisive break from the past since 1991. These included the abolition of the industrial licencing system (with a few exceptions), abolition of controls on capital issues, liberalization of the import licencing system, substantial lowering of tariffs on imports and adoption of a realistic exchange rate policy. Economic reforms in these areas had no doubt yielded positive results. The average trend rate of growth, by 2005, was close to 6 per cent per annum, the corporate private sector was showing signs of resurgence, access to domestic capital markets had become easy and, above all, the BOP and the reserves position had become strong. The point of drawing attention to certain other important areas where the deadweight of the past was holding back policy reforms was simply to highlight the fact that further acceleration of the growth rate and higher public investment in areas which benefit the poor (for example, public irrigation, rural infrastructure, literacy and healthcare) was unlikely to occur until the old mindset changed.

## The Power of Distributional Coalitions

At the time of Independence, India's feudal past, large-scale poverty, vast differences in the distribution of income and wealth and its divided social structure created doubts whether its unity as a nation and its democratic experiment with adult franchise would last very long. Some of these doubts were further reinforced by the violence during the Partition and later the differences among the states over official languages and financial devolution. However, after India successfully went through three or four general elections, and the country's federal system became politically viable and generally accepted across the country, India's democratic experiment won worldwide admiration.

A number of theories have been advanced to explain the reasons for India's survival as a democracy despite its many internal contradictions. The research findings and various hypotheses advanced by scholars are certainly useful in understanding the nature of India's democracy and the various forces at work in determining its evolution. However, in a functioning democracy like India's, where the vast majority of voters is poor, it is still difficult to understand why policies that do not benefit them continue to enjoy so much political support.

While there may be a number of alternative explanations for this state of affairs, perhaps the most convincing answer is to be found in the observed fact that, as highlighted in the public choice theory, the political decision-making process on economic issues in most democracies is driven by special interests rather than the common interests of the general public. These special interests also happen to be more diverse in India than in other, more mature economies. There are special regional interests not only among the states, but also within the states depending on the electoral strength of the party in power. Economic policy-making at the political level is further affected by the occupational divide (for example, farm vs non-farm), the size of the enterprise (for example, large vs small), caste, religion, political affiliations of trade unions or the asset class of power-wielders and a host of other divisive factors. As a result, most of the economic benefits of specific government decisions are likely to flow to a special interest group or so-called 'distributional coalitions'. These coalitions are always more interested in influencing the distribution of wealth and income in their favour rather than in the generation of additional output, which has to be shared with the rest of society.

Also, the delivery of government benefits to special groups

gave rise to a whole process of bargaining and conflict resolution among various interests. As a result, a large number of middlemen emerged across the political spectrum. Further, as elections became more expensive and more frequent with an uncertain time period during which funds could be collected in different states, there came to be greater tolerance of political corruption as an unavoidable feature of the electoral process.

Thus, contrary to what was envisaged by the founding fathers of the Republic and the vision of planners, in several crucial areas of political-economy, balance in actual practice turned out to be narrow and wasteful. How did the stranglehold of special interests last so long? Where were the majority of people who did not gain sufficiently from the economic bargaining process? The answer is not difficult to find. The simple fact is that the so-called 'majority' was fractured into a large number of subgroups of individuals who were divided among themselves by several factors (such as caste, religion, location and/or occupation), while special interests were united in protecting their share of the economic pie. This was really why the so-called 'haves' were so much powerful than the 'have-nots' in the society. It were, for example, the trade unions of employed persons (or the 'haves') that were likely to go on strike when their economic interests were threatened, rather than the vast majority of the unemployed (or the 'have-nots) across the country.

At this point, it must be made clear that the important role of special interests in determining political economy outcomes is not an argument in favour of unfettered free markets, or the need for an economy without government regulations and laws. The issue here is not 'market vs government'. The problem with the Indian economy was not whether the market was free or not, but that its freedom was in the wrong domains. It was common knowledge that in most parts of India, government permissions, regulatory approvals or licences could be obtained at a price. In these domains, the problem was that of excessive marketization.

On the other hand, in other areas where the market ought to be more free (for example, the labour market or international trade), India was strapped in bureaucratic red tape.

Two more caveats were necessary when considering the power of dominant coalitions in determining economic policy outcomes. The point was not that these coalitions always emerged as winners in determining the direction of public policy, or that all politicians pandered only to special interests. There were honourable exceptions, and there certainly were leaders who gave primacy to the general interest, but they were likely to be exceptions rather than the rule. They were also likely to face considerable hurdles in successfully pursuing economic policies which adversely affected the special interests of the organized groups. Similarly, there were situations (such as war, natural catastrophes or religious conflicts) when a unity of purpose emerged among all sections of the people to promote the common good.

It is the power of special interests and dominant coalitions which explains why, in India, policies and programmes that benefit only a small proportion of the poor command such substantial support among all political parties. Thus, for example, the creation of additional government jobs at two or three times the market wage rates is the favourite preoccupation of most ministers in charge of administrative ministries and public enterprises.

Over time, the expansion of the government's salary bill had fiscally disempowered most states (and the Centre), leaving them with very little capacity to undertake capital expenditure to improve facilities and public services for the vast majority of people. Yet, despite the huge fiscal drain, the number of government jobs in relation to the size of a state's population was relatively small and benefitted only a handful of persons in the state capital. Their unions were also the most powerful in determining the government's expenditure priorities. The same was true of the passionate advocacy of job reservations for particular sections of the people. Only a very insignificant percentage of the population belonging to a particular reserved category actually benefited, but the political noise about the benefit of such reservations for the disadvantaged and the poor was enormous. The point here was not whether reservations were desirable or not. There was undoubtedly a good case for them. However, it was striking that the dominant political coalitions seemed to be more interested in the distribution of the few available jobs rather than in increasing overall output, employment and the general welfare of the people.

An important area of the economy that had been severely affected by fiscal disempowerment in the states had been that of agriculture. Since 1994–95, the rate of growth in agricultural production had been down to less than 2 per cent per year compared to over 4 per cent earlier (1980–81). With the exception of a couple of years, monsoons had generally been good. However, the availability of public irrigation and power to the average farmer had deteriorated because of low investment, poor maintenance and administrative apathy. An important reason for this state of affairs was lack of financial resources with district authorities, including local panchayats, for capital investment or maintenance. The decline in agricultural growth rate was the primary cause of high levels of poverty and increasing disparity in income growth in sectors which were dependent on public investment. The need to reduce subsidies which do not benefit the poor and increase public investment seldom figure in the public debate among political parties or political leaders, even though areas with agriculture as the primary occupation send the largest number of representatives to parliament and state legislatures.

## The Disjuncture between Economics and Politics

While democracy had clearly spread to the remotest area of the country in ever-widening circles of political awareness among hitherto subordinate groups, the political system had also become increasingly unresponsive to the economic interests of the median number of the poor, disadvantaged groups. Politicians were seldom penalized by the electorate for their endemic poverty or the erosion of the public delivery system. There were very few assurances that commitments made by one government (or leader) will be kept by successive ones, or even by itself when under pressure. A political party that introduced some reforms (for example, disinvestment in privatization of public enterprises) was likely to be quick to oppose them when it was no longer in power.

Barring a few occasions when a particular issue acquired overwhelming national importance, such as the Emergency or India's role during the war in Bangladesh, there was also no discernible pattern that showed why particular parties or candidates win or lose elections. Until the elections in 1977, after the Emergency, the Congress party returned to power, with varying majority, at the Centre and in most of the states, irrespective of the economic results of policies pursued by it. India had its first severe foreign exchange crisis in 1956, and it became increasingly dependent on official aid from abroad for more than 35 years after that. Yet, a staple of the economic agenda put forward by the government and the Planning Commission over these years was a clarion call for 'self-reliance' and control over non-official foreign capital inflows. The removal of poverty through government control over investments and an expanding public sector were also important parts of the economic agenda. The anti-poverty objective found a particularly powerful expression in the slogan *Garibi Hatao* (Remove Poverty) during the 1971 elections. The government, which had already been in power for 24 years without being able to remove poverty, won the elections again with a massive majority on the basis of this promise!

The growth rate picked up after 1980, when the Congress party returned to power after three years of a Janata Party government. However, it lost the elections again in 1989 for reasons which had little to do with the economy (although by that time, an external crisis was brewing because of excess external borrowings and rising fiscal deficits). The Congress party came back to power again in 1991 in the midst of one of the worst economic crises, and launched a programme of economic reforms that was universally acclaimed. The crisis was soon over and India became externally strong. However, the party lost the next elections in 1996.

The chronicle clearly establishes the absence of economic performance as a factor in determining electoral outcomes. Economic considerations seem to be even less important at the state level. In the two largest states, Bihar and Uttar Pradesh, different parties with no credible record of economic performance or pro-poor policies were returned to power time and again. As for individual MPs, a report based on returns filed by candidates in the 2004 Lok Sabha elections showed that as many as 100 members (in a House of 543 members) had criminal charges against them. In addition to various other crimes (such as murder, fraud or kidnapping), almost all of these members had charges of financial corruption against them. Six of these members went on to become ministers in the Union cabinet. The previous Lok Sabha also had a sizeable number of members who had criminal charges against them. Thus, the economic progress and success or failure in removing poverty have very little effect on electoral outcomes despite the overwhelming attention paid to these objectives in party manifestos and election campaigns.

It may be recalled that during the freedom struggle, an important unifying political force across the country was the desire to break away from the impact of colonial economic policies which had kept India poor and stagnant. In the post-Independence period, for the first 15 years or so, the Five-Year Plans had raised high hopes of India becoming an economic and industrial power in the foreseeable future. In addition to his political charisma, Pandit Jawaharlal Nehru's initiative in launching India's ambitious Plans and expectations they generated in all sections of the people was an important reason for the massive electoral mandates in favour of the Congress during the first three general elections after Independence. However, after one of the worst droughts in 1965, it became increasingly clear that India's development strategy and economic policies were not yielding the results and benefits claimed for them. While the electoral rhetoric still proclaimed the supreme importance of anti-poverty programmes and self-reliance, the economic record of the government in achieving these objectives became less and less important over time.

Apart from economic growth, and foreign policy and defence, there were generally no other national issues of similar importance which could play a dominant role in determining electoral outcomes across the country. It was not surprising, therefore, that gradually sectarian, local and regional issues began to play an increasing role in determining electoral outcomes. This explains the increasing importance of regional parties and the reason why the electoral verdict became more divisive and fractured across the country. It also changed the shape of the electoral agenda and the role of political parties and their leaders during elections.

Sectarian and local issues, when seen in the national context, naturally become more divisive. If there is a government of a different party or parties at the Centre, the lack of development and continuing poverty in a particular state can always be said to be due to inadequate assistance from the Centre. Similarly, there is much greater stress on reservations or the distribution of existing jobs, or the extension of special benefits to particular castes or sections of people, areas or occupations. The nature of the accountability of regional parties for performance in their states as well as of their leaders also changed. They were not accountable for the failure to deliver what they promised or for the generation of employment or incomes for the poor, as some other party or some other centre of power could always be blamed for the continuing poverty and lack of progress.

At the Centre, accountability for actual performance also tended to be weak. Fiscal accountability became further and further removed, as a failure on the fiscal front could always be attributed to the actions of the previous government or to the unavoidable compulsions of higher growth rate. Big promises for poverty alleviation or employment generation were made, but it was expected that a new government would have taken office by the time the actual results of these policies became evident. In addition to political apathy and the lack of accountability for economic performance, there was the growing ineffectiveness of most institutions—courts, bureaucracies and the police.

To conclude, there were several powerful forces at work, which could adversely affect India's economic performance in the future. The actual outcome could not be up to India's high potential unless there was a significant change in policy perceptions. There was increasing disjuncture between politics and economics. The power of special interests in determining policies of the state had increased, there was growing fiscal disempowerment and governments were less accountable for the outcome of policies initiated by them. While the vigour and varieties of India's democratic politics is a matter of comfort and joy, its lack of success in serving the real economic interests of the vast majority of the people, particularly the poor, has been a matter of profound concern.

# 12

## A DEFINITIVE AGENDA
## FOR POLITICAL REFORMS

### 2018

The present government, with a majority, has the ability to bring some important reforms not only in economic policies but also in respect of the working of the political and administrative system, which can deliver public services to all the people with least diversion, delay or multi-tier corruption in allocation.

It hardly needs to be emphasized that a fundamental 'systemic' change that dominated the working of India's politics over a long period before 2014 was the fragmentation of political parties. As a result, India had as many as nine governments in the past 25 years—with an average life of about two and a half years. Of these, only two coalitions survived their five-year terms. Excluding these two full-term coalition governments, the average term of seven governments—with enormous powers to allocate resources, control public enterprises and decide interstate allocation of investments—was less than two years. The crucial point, in view of past experience, is that at the time of formation of a coalition government, the general expectation of small and regional parties was that the enormous powers that their nominees, as ministers, enjoyed may not last very long or that it may change if a more powerful leader of one or two large parties in coalition so decided.

The proposed agenda for political reforms, outlined in this essay, is necessarily selective and not comprehensive. The proposed changes to make the present system more accountable and strengthen the democratic process would be difficult to accept or implement because of the inherent conflicts of interest among different sections of the political spectrum. However, this is a minimum—and practical—agenda that deserves consideration and debate in legislative bodies, the media and other institutions of the civil society.

## A Federation of States

Articles 245 to 255 of the Constitution of India deal with the distribution of powers between the Union and the states. The Centre has exclusive powers to make laws in respect of matters enumerated in the Union List (such as defence, foreign relations and financial matters concerning the whole of India). The states, on the other hand, have exclusive powers to make laws in respect of matters enumerated in the State List. These generally include matters where uniformity across the different states in respect of legal and administrative matters is not considered necessary (such as internal law and order, agriculture, and trade and commence within a state).

There is also the Concurrent List under which both the Union and states can make laws. These include matters where the Centre can make laws applicable to all of India, but where individual states are also entitled to pass laws of specific interest to them. The residual powers, i.e., powers to make laws on any subject that is not listed in any of the above lists, rest with the Union (unlike certain other federations, such as the US, where the residual powers lie with the states). The Centre also has the powers to make laws that are applicable to two or more states, if the concerned states so request, on a matter listed in the State List.

This scheme for the distribution of powers between the Union and the states has stood the test of time, and is a tribute to the foresight of the framers of India's Constitution. In a country with such great diversity in languages, religions, castes and levels of development, this scheme also proved to be a major unifying force. All states are represented in the two Houses of the parliament and work together on the treasury benches or in the Opposition. Regional issues and matters of interest to particular states are open to discussion in parliament, and are generally resolved through a consensus.

There are, of course, long-standing interstate disputes (particularly on water or sources of energy), which flare up from time to time. However, even these have not threatened the unity of India because of the Union's conciliatory role and the representation of most states in the Union Cabinet. An outstanding initiative taken on 1 July 2017 by the present government is the introduction of Goods and Services Tax (GST) with mutual agreement between the Centre and the states.

In the context of recent political developments at the Centre, and the emergence of multi-party coalitions of different types as a regular feature of governments, it is perhaps necessary to also review the present division of powers between the Union and the states. In view of external terrorist linkages and other factors, there is an urgent need to consider transfer of powers for the maintenance of internal security to the Centre from the states.

In the economic area, it is desirable to consider a reverse transfer, i.e., powers and responsibility for financing development programmes should be transferred from the Centre to the states. At present, the states formulate their plans, but the responsibility for the approval and provision of sufficient resource for implementing them rests with the Centre. An investment plan for a particular period may be launched by one multi-party coalition government in a state, and approved by another combination of parties in power at the Centre. However, thereafter there may be a change in the government in the state and/or at the Centre. With every change in the government at the Centre, there will be a change in the composition of the earlier Planning Commission (which is now the NITI Aayog [also known as National Institution for Transforming India]).The same is true at the state level. There is a realignment of the political relationship between the parties in power at the Centre and in different states with a change in government. Governments are political bodies and their decisions are discretionary. Thus, it is likely that each year, the actual flow of central assistance to different states will be increasingly determined by the timings of elections and the party composition of governments at the Centre and the party/parties in power in different states.

It is desirable to take action on two points. First, more financial powers and increased responsibility for the implementation of development programmes should be entrusted to the states. This is not because all states are likely to be more scrupulous or consistent in the exercise of their powers, but because greater transparency and competition among states would, at least, ensure that the better governed states have easier access to financial resources and the opportunity to implement their programmes.

Just as the Finance Commission is constitutionally empowered to decide on the division of tax resources between the Centre and the states, a similar federal commission should be statutorily set up to decide on the devolution of all other forms of central assistance. The allocation of non-tax central assistance should be related exclusively to the implementation of approved anti-poverty and development programmes in physical terms. The greater the success of a state in implementing a programme in relation to its target in quantitative terms, the higher should be the allocation of central funds to that state.

Second, all appointments in autonomous institutions, regulatory bodies, public enterprises, banks and financial, educational and cultural institutions in the public sector should be entrusted to specialized bodies set up on the same lines as the Union Public Service Commission (UPSC). These bodies should follow transparent procedures and criterion for recommending appointments to top positions. Their recommendations should be invariably accepted by the government (as is the case with UPSC recommendations for entry into the civil services and other appointments under its purview). Similar procedures, at an arm's length from the government, may be adopted for top appointments in the services. Recent developments and the controversies surrounding them in many of India's top institutions, highlight the need for urgent action to insulate public institutions from excessive political interference in their day-to-day work.

## The Council of States

In respect of elections to the Council of States (i.e., the Rajya Sabha), the Representation of the People's Act, 1951 was amended by the parliament in August 2003. The two significant amendments were: (i) persons elected as members of the Rajya Sabha do not have to be residents of the states that elect them; and (ii) the secret voting procedure, which is applicable in all other elections, has been replaced by open voting. A dissenting member who votes against a party candidate is likely to be removed from his/her party as well as from the state legislature for indiscipline and defection.

At first glance, the above two amendments appear quite reasonable. It is a well-known fact that in the past, some members who were not ordinarily residents of a state, had declared themselves to be residents of that state in order to qualify for elections to the Rajya Sabha. Similarly, it has also been noticed that some electors in state legislatures had voted in favour of candidates belonging to other parties in exchange for financial and other favours.

In practice, however, the combined effect of the above two amendments has had a substantial impact on the composition of the Rajya Sabha and made the so-called 'Upper House of the Parliament' even less representative of people than was the case in the past. Neither the people of a state directly, nor their representatives in the state legislature indirectly, have any voice or discretion in electing their representatives to the Rajya Sabha. The choice of members now depends entirely and exclusively on the leaders of various parties. Anyone with sufficient resources, organized manpower and access to leaders can be elected from anywhere depending on the ability of a party to swing sufficient number of votes in the state legislature. A system of *quid pro quo* among the parties has also developed where one party can provide balancing support to another party in one state in exchange for similar support by that party in another state.

Over a period of time, the Rajya Sabha is also likely to become a safe haven for leaders who fail to get elected to the Lok Sabha. Public-spirited individuals and those with a background of service to the people of a particular state will still get elected as nominees of different parties but, over time, such cases are likely to become fewer in number. Taking into account the working of our system of parliamentary democracy, with more than 50 parties being represented in state legislatures and the parliament, there is no doubt that, eventually, the composition of the Rajya Sabha would become vastly different from what was originally envisaged in the Constitution.

A reform of the present system for elections to the Rajya Sabha is now urgent. In case it is not politically feasible to reform the electoral process, it would be much better for the functioning of our democracy to have a unicameral parliament, as is already the case in some states such as Arunachal Pradesh, Assam, Chhattisgarh and Delhi. This will not only save time and budgetary resources, but will also prevent further erosion in the federal character of the Indian Union.

One argument against having a unicameral legislature is that it may affect the quality of the Cabinet. This may happen in case some prominent and qualified persons belonging to a political party (which is called to form the government) happen to lose the Lok Sabha elections or are not inclined to contest elections. At present, they can be elected to the Rajya Sabha and are able to find places in the Cabinet, if their party so desires. The same is true of some highly qualified persons from different professions, industry and trade.

These concerns are valid. It is in the larger public interest for the PM to be able to appoint the most qualified persons in the country to the Cabinet. However, this objective can be achieved by adopting a constitutional amendment to the effect that a majority government can, if it so wishes, appoint, say, up to 25 per cent of members of the Cabinet from outside the parliament. Those who are appointed in this manner may be authorized to participate fully in the proceedings of the Lok Sabha in their ministerial capacity without having the right to vote. Interestingly, this is precisely the case now in respect of MPs who are appointed to the Cabinet.

## State Funding of Elections

The issue of the state funding of elections has been considered from time to time in parliament and other forums. However, so far, no consensus has emerged even though there is general agreement that the need to collect large funds for elections is a primary cause of political corruption. It is also a known fact that, over time, while large amounts of funds are being raised in the name of political parties, a substantial portion of such funds are being diverted for personal use. The print and electronic media have exposed several high-profile cases of the accumulation of illicit wealth by chief ministers, ministers and other leaders in and out of office.

It is obvious that given the large number of persons with criminal records who are active in politics, state funding of elections will not eliminate corruption entirely. However, it would at least help those who want to remain in politics without having to indulge in corruption. State funding may also help in reducing the acceptability of corruption as an unavoidable fact of Indian political life and strengthen NGOs or other individuals in their fight against corruption.

One argument which is frequently advanced against state funding is that it would favour large parties and would, therefore, be unfair to small or new parties. Another argument is that the fiscal cost of such funding will be high and unbearable for many states as well as the Centre. While there is indeed some merit in both arguments, they are, by no means, persuasive and compelling.

To take the second argument into consideration first, the size of the budget expenditure by the central government is estimated to be more than ₹20 lakh crore for financing Lok Sabha elections and for providing some support to state governments for the state elections. Even if such elections are held twice every five years (because of greater political instability), this amount should be sufficient to provide adequate funds for legitimate electoral expenses in each Lok Sabha constituency. In terms of the central budget, the amount to be earmarked for elections could, thus, vary between 0.2 and 0.4 per cent of the total expenditure annually (depending on the frequency of elections). By no means can this be regarded as an unbearable fiscal burden for a cause as vital as election funding.

It may also be mentioned that the central government allocates nearly ₹3,500 crore annually to fund the Members of Parliament Local Area Development Scheme (MPLADS). State governments have similar schemes for their Members of the Legislative Assembly (MLAs). In view of the significant misuse of such allocations, as highlighted by the media from time to time, which has led to the expulsion of several MPs, it is desirable to discontinue MPLADS-type schemes both at the Centre and in the states. From the country's point of view, it would be much better if budgetary funds allocated for such schemes were used for election funding.

In addition, it is also feasible to introduce a practical scheme for the equitable distribution of electoral funds across all political parties, whether big or small. The proposed distribution formula, given below, is by no means perfect, but it should broadly meet the legitimate concerns of the small parties:

- Funds for elections to recognized political parties should be provided under two broad heads: (i) to reimburse certain categories of identified election expenditure; and (ii) to meet a relatively small amount of residual expenditure on staff and maintenance of party election offices.

- A predetermined category of actual expenditure, which should be eligible for reimbursement, could cover newspaper and television advertising for a specified period, say, two or three weeks prior to the elections and reasonable transport costs (by air and train) for election campaigns. Rules for reimbursement of actual expenses under these (and any other admissible heads) may be laid down by the ECI. Advertising by political parties may be limited to the amount which is actually eligible for reimbursement. In other words, parties that benefit from state funding for advertising, should not be allowed to spend any additional amount under this head, on their own account. They can incur additional expenditure on their own, if they wish to, on all other items, such as, transport, staff and offices.

- The division between 'big' and 'small' parties for the purposes of allocation of funds may be made according to a benchmark, approved by parliament after appropriate consultations with the ECI. Thus, parties with, say, a minimum of 10 or 15 per cent of the seats in the Lok Sabha or the state legislatures for (state elections) may be considered as 'big' parties; the rest can be considered as 'small' parties.

- Reimbursement of actual expenditure under the prescribed heads may be the same for all large parties (as defined by parliament for this purpose), and proportionately less for smaller parties (depending on the actual number of seats held by them in the Lok Sabha or state legislatures).

- Allocation of funds to meet residual expenditure on staff and offices may be weighted by the number of seats held by each party. The weights may be suitably devised to ensure that the larger the party, the higher is its entitlement for funds on this account. At the same time, small parties should not be subject to undue disadvantage.

As a large and vibrant democracy, India is not alone in facing the problems of electoral funding through legitimate means. The costs of contesting elections in India and elsewhere have increased phenomenally in recent years because of changes in the methods of communication. In addition to personal campaigning, the use of electronic media for sending messages to voters has become unavoidable in all democracies. While costs have increased, and there are several elections to fight every year at different levels (i.e., national, state and district), contributions from reliable sources have dwindled. Fewer persons now become party members or make contributions to political parties. The same is true of charitable organizations and the corporate sector.

In order to overcome these problems, some of the old as well as new democracies, including the UK and the US, have introduced some funding of political parties through transparent and verifiable rules. India must do the same as early as possible.

A further step that deserves urgent consideration is that of state funding of some additional electoral expenses, as per certain guidelines which are in public interest. This will enable small as well as large political parties to avoid reliance on undeclared donations.

The Budget for 2017–18 announced the following measures for introducing transparency in electoral funding of political parties through donations by individuals, partnership firms, Hindu Undivided Family[3] and corporates:

- In accordance with the suggestion made by the ECI, the maximum amount of cash donation that a political party can receive will be ₹2,000 from one person;
- Political parties will be entitled to receive donations by cheque or digital mode from their donors;
- As an additional step, the RBI Act will be amended to enable the issuance of electoral bonds in accordance with a scheme formulated by the government. Under this scheme, a donor could purchase bonds from authorized banks against cheque and digital payments only. They will be redeemable only in the designated account of a registered political party.
- Every political party would have to file its return within the time prescribed in accordance with the provisions of the Income-tax Act.

These steps are certainly worthwhile, as exemptions to the political parties from payment of income tax would be available only subject to fulfilment of these conditions. At the same time, it is also likely that the above provisions would not be sufficient to meet actual electoral expenses being incurred by most political parties at the Centre as well as states.

---

[3]Hindu Undivided Family (HUF) consists of all persons directly descended from a common ancestor, and also the wives and daughters of the male descendants. For instance, you and your spouse along with your two children can create an HUF and get certain relaxation in computation of taxes. (https://www.business-standard.com/about/what-is-hindu-undivided-family, accessed on 23 March 2021).

## The Role of Small Parties in the Government

In a democratic state, every citizen has a legitimate right to vote, contest elections and launch a political party. If a party enjoys the minimum electoral support prescribed by the ECI, it is recognized as a national or regional party. At the national level, regional parties are also eligible to join a coalition government. Irrespective of the size of their representation in the parliament, they can continue to function as separate parliamentary parties with their own agenda. In principle, this is a reasonable arrangement in a diverse, multiparty country with a federal constitution like that of India.

An important principle of parliamentary democracy is that the government is formed with the support of a majority of directly elected members in the House of the People (i.e., the Lok Sabha). Each member has a single and equal vote. Once a government is formed, whether by one party or by a number of parties in a coalition, it is supposed to be collectively responsible to the parliament and, through it, to the people of the country.

Unfortunately, over time, these fundamental principles of parliamentary democracy have been compromised. Small parties, with less than 5 per cent of the national votes and an even smaller number of MPs, now command a disproportionate influence as partners in a coalition government. Even a collective decision of the Cabinet can be shelved or overturned at the insistence of a supporting regional party. If the coalition government consists of a number of small parties and is also dependent on the support of other parties outside the coalition for survival, then the situation becomes even more complicated. The government may continue in office, but it is unlikely to enjoy sufficient political authority for efficient governance.

Another consequence of the formation of governments with the inside and outside support of a large number of small parties is the possibility of political instability. After 1989, as many as six governments were unable to complete their full term. This was also the case in respect of two coalition governments which were briefly in office during the period 1977–79 (after the Emergency of 1975 was lifted). It is also worth recalling that two of the worst economic crises faced by India, in 1979 and 1990, were, in no small measure, due to the inability of governments then in power to take timely and appropriate corrective action because of uncertainty about their survival in office.

Political uncertainty and instability are sometimes unavoidable in a parliamentary democracy, where a large number of parties have conflicting interests. Although India now has a majority party in office, in future, it is important to ensure that all parties that form a coalition function collectively to provide efficient public administration. To this end, it is particularly desirable to reduce the disproportionate power enjoyed by small parties that decide to join a coalition.

At present, a small party is free to join a coalition and hold crucial ministerial berths. It is also free to follow its own regional or sectoral agenda in the exercise of its ministerial responsibility. In case it is dissatisfied with a decision of the Cabinet, it can threaten to walk out of the coalition and reduce the government to a minority. In order to avoid fresh elections, it can continue to support the government from outside if and when a no-confidence motion is moved by the Opposition. In this way, the members of a defecting party can continue in parliament/legislatures. At the appropriate time, they can also join another coalition.

In order to prevent defections by individuals or groups of members of a party in the parliament/legislature, in 1985 and again in 2003, the Constitution was amended to disqualify them from continuing as members or holding any other public office until their re-election. A similar measure should be introduced to disqualify members of a party (with say, less than 10 or 15 per cent of seats in the Lok Sabha, as may be decided by the parliament) who opt to join a coalition and then decide to defect. It should be made mandatory for all members of such a party to seek re-election.

As provided in the 2003 amendment, members of a defecting party should also not be permitted to hold any public or ministerial offices during the remaining part of the term of parliament/ legislatures. There is no justifiable reason why members of a small party should be put in a more favourable position than any other group of defecting members. Indeed, it can be argued that the present system provides a built-in incentive for the fragmentation of a large party into smaller parties at the time of elections. The leader of a small party enjoys all the benefits of being part of a larger party formation (e.g., occupying a ministerial berth) without any of its disadvantages. These rules should also apply to 'independent' members who opt to join a coalition government.

Further, in order to reduce the threat of political instability in the future, it is also desirable to introduce an amendment in the rules of business in the parliament. It may be provided that all parties in government should become members of the same parliamentary party under the banner of their coalition. They should not be recognized as separate parties for purposes of parliamentary business. Thus, for example, the NDA or the United Progressive Alliance (UPA) or any other coalition that forms the government should be considered as the NDA or the UPA parliamentary party, respectively, as long as it is in office. Such an amendment will have the salutary effect of formally recognizing all parties in a coalition as a joint parliamentary party for conducting the business of government in the parliament/ legislatures. All such parties may, of course, continue to have their separate identity for all other purposes, including the power to nominate their own candidates during elections.

## Reform of the Government

In India, governments at the Centre and states, along with their agencies, have, practically, unlimited powers to pass laws, notify rules and regulations and determine economic and social priorities. Some of these may require parliamentary or legislative approval, but as long as a government has the requisite majority, such approval is a formality. While available powers are enormous, it is also a fact that the authority of the government to actually enforce laws and rules is minimal. Part of the reason is, of course, judicial delays.

Several surveys and opinion polls have provided telling statistics about the extent of corruption in government agencies. A survey by the Public Affairs Centre, based in Bengaluru, found that in recent years, every fourth person in one of the large cities in India ends up paying a bribe when dealing with agencies involved in urban development, electricity, municipal services and telephones.

Interestingly, despite the deepening crisis of governance, in 2017–18, India was witnessing a new growth momentum. This paradox could largely be explained by three factors. First, beginning in the early 1980s, the government's heavy-handed control of the non-governmental sectors in manufacturing as well as services, was lifted. A second factor was the gradual opening up of the economy through reduction in protective tariffs and the abolition of import and export quotas of various types. India became an attractive global destination for capital, skills and business outsourcing. Finally, particularly after the 1997 Asian crisis, India managed its external sector exceedingly well. After nearly four decades of periodic crises (beginning in 1956), India emerged as a country with a strong BOP and one of the highest levels of foreign exchange reserves.

It is interesting to note that all the three factors mentioned above were related to positive changes in macroeconomic policies, which created a more competitive environment, and removed extensive governmental controls over individual and corporate initiatives. These had very little to do with institutional or micro-level changes in the administrative and governance structure within the government.

The fact that the overall growth rate in the economy accelerated in the recent period due to resurgence in the private sector, makes the need for reforms within the government even more urgent. In a poor country with a large population where the bulk of the people are dependent on agriculture and have access to a few basic amenities, a high growth rate in the national income by itself cannot reduce disparities or eradicate poverty. Government intermediation in favour of a more equitable distribution of the benefits of growth through the provision of public services and public investment in basic infrastructure is essential.

Even if it is estimated that as many as 200–250 million people are currently benefiting from the high growth rates in manufacturing and services in the private sector, more than 800 million persons in India still continue to be at the periphery of prosperity for quite some time. Meanwhile, the widening of disparities among different sections of the people can cause severe strains in the political and social life of the country.

The kind of reforms which are required and are feasible in the current political scenario is a debatable topic. There are, however, a few vital principles which need to be adopted as early as possible in order to guide the process of reforms in the next few years:

- The political role of the government in the economy needs to be redefined and prioritized. At the macroeconomic level, the political (i.e., ministerial) role of the government should be to ensure a stable and competitive environment with a strong external sector and a transparent domestic financial system. While the macroeconomic priorities (for example, the trade-off between growth and inflation) may be decided by the government, the instrumentalities for achieving these objectives must be left to autonomous regulatory and promotional agencies.
- The government's direct role in economic areas must be reset in favour of ensuring the availability of public goods

(such as roads or water) and essential services (such as healthcare and education) to the people. In these areas, the government's role must expand substantially. At the same time, its role in managing commercial enterprises deserves to be correspondingly reduced. The latter objective should be achieved without, in any way, affecting the financial and other benefits of those who are presently employed.

• Another important priority is simplification of administrative procedures and reduction in the number of agencies, at different levels, involved in providing clearances for undertaking any activity. This is an area where the supply of corruption by public servants creates its own demand. Except in selected areas of paramount national interests (such as security and defence), it is desirable to cut through the elaborate red tape and rely primarily on 'self-certification'. The government can lay down standards and norms (for example, in respect of environmental impact or safety) and the entity concerned may be required to 'self-certify' that these have been complied with as per the notified procedures. Government agencies can make random checks and in case there are violations, appropriate penal action can be taken.

• A related area is transparency in the decision-making process within the government. A major step in this respect has been taken with the enactment of the Right to Information Act (RTI), 2005. A further step in this direction is to make it mandatory for all ministries and departments of the government to voluntary make information on the decisions taken by them available to the public (excluding security-related subjects).

• Case studies of international experience in the management of public services show that the objective of such programmes can be achieved better, and at less cost, if a distinction is made between the ownership of

these services (by the government) and the delivery of
such services (by NGOs and local enterprises). In such
cases, the public authorities retain the responsibility
for regulating and monitoring the activities, providing
subsidies where necessary and laying down distribution
guidelines. In India, a noteworthy example of public–
private collaboration in the area of public services is
the PCOs, which have revolutionized the availability of
telephone services across the country since the 1990s.

## Ministerial Responsibility

A minister, as the political head of a ministry, enjoys enormous
executive powers. Part of the rationale for entrusting politically
appointed ministers, of whom several have very little previous
administrative experience, is that the ministry is supposed to be
accountable to the Cabinet and the parliament, through them.

While the above system is sound in principle, in practice,
there has been substantial erosion in the ability of the parliament/
legislatures to hold ministers responsible, either collectively or
individually, for the decisions taken by them on behalf of their
ministries. In addition to the principle of collective responsibility,
which shields ministers from taking individual responsibility,
another reason why they are not held accountable is that most
subjects in economic area, which are of direct interest to the
public, are in the Concurrent or State Lists of business.

The central ministers are free to make pronouncements,
approve policy guidelines and set all-India targets, but the actual
implementation of programmes happens to be in the hands of
individual states. A familiar excuse given by central ministers for
their failure in meeting the targets announced by them is that the
states, and not the Centre, are responsible for the implementation.
The states, on the other hand, blame the Centre for inadequate
allocation of funds, inappropriate guidelines and approval delays

by one or more ministries at the Centre.

Assuming that political parties, civil society and enlightened people of the society are serious about removing poverty and deprivation, a new institutional initiative is urgently required for enforcing ministerial responsibility. This can be achieved only if the doctrine of collective and concurrent responsibility for all actions of the government is replaced by the notion of individual responsibility for project implementation. The doctrine of collective responsibility can continue to prevail for all other political purposes, including the continuation of a government in office.

Another area where immediate action is necessary is that of lowering the bar on political corruption. A lid has to be put on the tolerance levels of corruption, at least at the ministerial level. Persons who have been charge-sheeted for corruption, fraud and similar criminal offences should not be permitted to take the oath of office and function as ministers until they are cleared of all these charges by the courts. A special procedure may be set up to immediately hold and expedite court hearings in cases of persons who are proposed to be appointed as ministers.

## Depoliticization of Civil Services

The basic issue that needs to be tackled for improving the morale of the civil services is really that of the 'separation of powers', within the executive, between ministers and civil servants, especially in matters concerning postings, transfers and promotions. The greater empowerment of the civil service must go hand in hand with the greater accountability of civil servants for their performance and ethical conduct.

There are two statutory provisions, namely Article 311 of the Constitution and the Official Secrets Act (OSA), 1923, which require reconsideration. Widely misused, Article 311 provides comprehensive constitutional protection for a person holding a 'civil post from being reduced in rank, removed or dismissed from service'. The OSA provides protection to civil servants and ministers from being held accountable for any action that can be labelled as secret by them. The RTI has substantially reduced the power of civil servants to deny information to the public. There is no reason why the OSA should still remain valid.

## Fiscal Empowerment

An important step was initiated in 2003, when the Fiscal Responsibility and Budget Management (FRBM) Act was adopted. The objective, as defined in the Act, was to ensure sustainable fiscal management and long-term macroeconomic stability. The FRBM Act was further amended in 2012. The FRBM rules formulated in 2013 provided for a reduction of gross fiscal deficit to 3 per cent by 31 March 2017. The budgets for 2017–18 and 2018–19 have exceeded the target of 3 per cent but deficits are still relatively lower than was the case earlier—the deficit was 3.5 per cent in 2017–18; estimated to be 3.3 per cent in 2018–19. On the whole, there is no doubt that as far as fiscal responsibility is concerned, India has done extremely well compared to most other emerging market economies.

The fundamental issue that requires attention relates to the present pattern of government expenditures and how these resources raised through revenue or fiscal deficits are being put to use. If deficits were used productively and could generate a sufficient rate of return to cover the repayment of past debt, the precise level of the deficit—within certain sustainable levels—would not have mattered all that much. Over the years, the basic problem in India has been that the bulk of government expenditure is devoted to the payment of salaries. New programmes are launched but governments are fiscally 'disempowered' from carrying them out in time. Despite large deficits, sufficient resources are not available for financing essential capital expenditure, improving public services and undertaking even the routine maintenance of infrastructure. Fiscal disempowerment is not confined to rural areas; even the fastest-growing cities are affected by the government's inability to provide civic amenities.

The same is true of several other states in India. According to official statistics, despite sharp increases in resource transfers from the Centre and high revenue deficits, development expenditure, as a proportion of total expenditure of state governments, has actually been declining. It is even more striking that, within the total development expenditure on all accounts, social, sector expenditure (comprising social services, food storage, rural development and warehousing) is also likely to show a proportionate decline (along with a rise in population).

In order to improve the economic conditions of the bulk of the country's population and reduce disparities in access to essential services, it is imperative for all states to take urgent measures to fiscally empower themselves. As past experience shows, higher fiscal deficits or larger transfers from the Centre do not provide an adequate solution to the problem of fiscal stringency.

## Legal and Judicial Reforms

Judicial delays in India are now legendary. In view of the long delays and multiple levels of appeal available to any person or organization, filing a case has become a convenient way of avoiding a contractual obligation or conviction for a crime. As a result, all courts, particularly high courts, are now overburdened with pending cases.

There are multiple causes for this state of affairs in a vibrant democracy like India. A person is free 'until proven guilty' and the burden of proof lies on the prosecutor. In principle, this is as it should be. Unfortunately, in practice, there are enormous delays at the level of investigating agencies in collecting evidence. Corruption among witnesses and others is also widespread.

An important reason for judicial delays is the plethora of legislative provisions on all aspects of national life, some of which are almost a hundred years old and internally contradictory. All ministries of the government, at the Centre and in the states, are keen to introduce fresh legislation and amendments to old statutes every time parliament/legislatures meet.

It is also an age-long practice, since the British times, for all bills passed by parliament/legislatures to include an omnibus provision that gives unfettered right to the government to notify 'rules' notwithstanding any other provisions of the Act or any other laws in force. The rule-making provision, which has the force of law, provides sufficient scope for the discretionary and arbitrary exercise of power by the executive.

Fortunately, a few states have successfully taken action to reduce cases pending for over 10 years in lower courts to less than 1 per cent of total pendency. Some of these states have also fixed annual targets and action plans for judicial officers to dispose of old cases and criminal cases where the accused have been in custody for over two years. In the next five years, it would be desirable for the Centre to reform the legal system in all the states to reduce pendency below 1 per cent. To achieve this objective, there is an urgent need to reduce the scope for appeals, adjournments and frequent hearings at different levels of the judiciary.

These political reforms which may be introduced by the government in the next few years is, by no means exhaustive. The implementation of these programmes would certainly increase political stability, reduce the powers of multi-party coalitions at the Centre as well as in states and help in reducing economic disparities in the long run.

# INDEX